ELEMENTS

of

ELECTRIC AND MAGNETIC

CIRCUITS

DAVID VITROGAN
Chairman, Department of
Electrical Technology
School of Science and Technology

Pratt Institute

TABLE OF CONTENTS Page

iv

CHAPTER 1—FUNDAMENTAL CONCEPTS
OF ELECTRICAL CIRCUITS

1.1 Energy

Engineering technology concerns itself with energy, that is the capability for doing. The most common forms of energy are mechanical, thermal, chemical, electrical and radiational. Some of our most important sources of energy are chemically bound. Such substances include coal, oil, natural gas and wood. Chemical reaction releases the energy in them and it is usually transformed into heat energy. The energy of a waterfall, of wind and of the tide is all mechanical energy. Mechanical energy manifests itself as potential or kinetic energy. Thus, when a weight is lifted above the ground it is capable of driving a nail or colliding with another object simply by falling. A body may therefore possess potential energy by virtue of its position. While the weight is lifted work had to be expanded and the potential energy may be looked upon as stored energy. When it is falling it gains speed and the energy due to its position, potential energy, has been changed into energy of velocity or kinetic energy. Thus the potential energy was not destroyed when the weight was falling but was converted into kinetic energy which is capable of doing work when the weight strikes the ground.

A body may also possess energy by virtue of its temperature. It is said to have thermal or heat energy. When a meteor burns away because of air friction upon entering the earth's atmosphere kinetic energy is being converted into thermal energy. Heat engines are devices which convert heat energy into kinetic energy. Most of the energy with which engineering is concerned is traceable to the sun and reaches the earth in a form of radiation or radiant energy. As an example some of this radiant energy appears as heat on reaching the earth, evaporating water from the oceans and lakes. The water vapor then rises, condenses in the cooler upper atmosphere and falls as rain or snow; some of it eventually finding its way into the reservoirs of hydroelectric plants. Hydraulic turbines are used to convert the potential energy stored in the water into mechanical energy. The mechanical energy in turn, is then transformed into electrical energy by electric generators.

At the present time we do not really know the source of the radiant energy from the sun. We are quite sure that energy can be created at the expense of matter, a process that goes on spontaneously in radioactive chemical elements continuously. The process can not be arrested but it may be accelerated, as in the atomic explosion, to create tremendous amounts of energy in a very short time. It is hoped that in the near future energy created from matter will become most useful in engineering. Our present technology continues to depend on the conversion of energy from one form to another. We may therefore at present continue to rely on the law of conservation of energy. This theory states the total amount of energy in any system is fixed. Generally there are two types of conversion. Mechanical energy can be converted into heat energy by friction, say two blocks of wood are rubbed together. This is strictly a one way affair since the blocks will not move as a result of the heat which is applied. Such a process we call a non-reversible conversion. The second type is the reversible type which works both ways. In compressing a gas we find that the temperature of the gas rises indicating the conversion of the mechanical energy into heat energy. If we allow the compressed gas to expand suddenly its temperature falls indicating that thermal energy has been transformed into mechanical energy. Thus the conversion operates backwards as well as forward.

Energy can be transported or transmitted in nearly all of its forms. The choice is usually made on the basis of the costs, convenience and use to which it is put. Electrical energy can be regulated and controlled much more easily than most of the other forms. One of the particular advantages of the electrical energy is the utilization of this form of energy at remote points from its generation. Electrical energy can be transmitted over large distances readily and economically. The generation of electrical energy requires certain apparatus for the conversion of other forms of energy into electrical energy. The control and transfer of energy also requires electrical apparatus. Generators convert mechanical, chemical or thermal energy directly or indirectly into electrical energy. In turn the apparatus which utilizes electrical energy converts it into mechanical energy for many industrial, communication or household uses. Such apparatus also converts the electrical energy into sound for the reproduction of speech and music; into chemical energy for electrochemical processes; into light for illumination and television;

into thermal energy for furnaces and miscellaneous heating devices, industrial and household.

In forming a practical electrical system, electrical generators and the apparatus for the utilization of the electrical energy, which are referred to as loads, together with control, measuring and other auxiliary devices are physically interconnected by electric conducting links, such as wires, or indirectly by electromagnetic means. Under certain conditions the system of interconnected apparatus or parts of the system may be looked upon as a physical electric circuit or electrical network.

The purpose of electric circuit theory is to provide a means of determining current and voltages throughout the electric circuit. By determining these quantities we may also find the associated electric charges, the magnetic flux linkages, power and the energy associated with the circuit. Circuit theory does not concern itself with the details of construction or the theory of operation of the apparatus that forms part of the physical circuit. Circuit theory makes use of symbolic idealization of circuit apparatus and hence involves abstractions more so than a subject that concentrates on the study of specific devices such as rotating machines, transformers, vacuum tubes etc. in which theory is closely related to the actual physical parts. Therefore in studying electrical circuits we must make a special effort from time to time to visualize the physical interpretation of the symbolic representation.

Electrical energy is dissipated in the form of heat in all electrical apparatus. Generally we concern ourselves with the time rate of energy conversion which is called power. There are two units of electrical energy in common use, the Joule and the Kilowatt-hour

1 Joule = 1 watt second
1 Joule = 0.738 foot-pound
1 Joule = 0.000948 Btu.
1 Kilowatt hour = 3413 Btu.

There are two units of electrical power, the watt and the kilowatt.

1 Watt = 1 Joule per second
1 Watt = 0.738 foot-pounds per sec.
1 Kilowatt = 1.341 horse power

The efficiency of an electrical device is defined as the ratio of the output power to the input power. Energy which is dissipated in heat in an electrical device usually represents a loss of useful energy. Therefore the highest efficiency means the least power dissipation in heat.

1.2 Electric Charge

Simultaneously with the development of the electric circuit concepts are the developments of the concepts of the nature of matter. The concepts of the atomic theory of matter are fundamental to all the physical sciences and technology. For an introductory study of the electric charge, the atomic structure of matter may be considered as composed simply of positive charges of electricity called protons, negative charges called electrons and uncharged particles, which are of the same mass as protons, called neutrons. A normal atom contains an equal number of positive protons and negative electrons. The charge of the electron and that of the proton are equal in magnitude but opposite in sign. Therefore, the normal atom is said to have no resultant electric charge.

The arrangement of the protons, electrons and neutrons in the normal atom is very orderly and in accordance with definite conditions for each particular element. The central core of the atom is called the nucleus and contains the protons and neutrons which are extremely massive with respect to the electrons. The electrons have very small mass and are contained in orbits or shells about the nucleus. The electrons of a single atom can not be separated from the nucleus without work being done upon them because of the attractive forces in the atom.

1.3 Conductors and Insulators

However in a large mass of a substance, there may take place a free interchange of electrons between adjacent atoms and a few electrons may be removed, even by friction, without expending a great deal of energy. The nature of the substance however influences greatly the relative ease with which the interchange of the electrons among atoms may be accomplished. Substances, such as metals, in which the interchange is relatively easy are called conductors. Substances in which this interchange is relatively difficult are called insulators. The large variety of materials that are used in the electrical industry differ in their ability to conduct electricity. As a consequence of the fact that in metals, at least one of the electrons in each atom is loosely held, a very small external force is needed to move or conduct some of the electrons from atom to atom in the metal. Electrical conduction that takes place in this manner is referred to as electronic conduction since the electrical current is transported by the loosely held or free electrons.

Atoms which have an excess of positive or negative charge, as a result of chemical action or bombardment with other atoms or electrons are known as ions. Conduction may also take place as a result of the motion of ions, that is charged atoms or groups of atoms. Such conduction generally takes place in solutions of acids, bases or salts or in fused salts. Generally, there is a chemical action involved and this type of conduction is known as electrolytic conduction. Other substances may conduct electricity by both means, by the electronic and electrolytic process. Gases in the conduction condition will generally contain free electrons and ions and hence will conduct electricity by both processes.

In those substances where there are very few free electrons available, referred to as insulators, very little electricity is conducted. Insulators are therefore very poor conductors of electricity.

In between the conductors and insulators are found a group of materials, which are destined to play a great role in the future in electrical technology. They are called semi-conductors and include copper oxide and selenium used in dry plate rectifiers, varistors and thermistors used in control instrumentation and transistors whose application in the field of electronics is unbounded.

1.4 Electric Current

Electric current represent the transport of electrical charge in a given time. In a metallic conductor, the free electrons are in a random, chaotic motion at all times, moving with very rapid velocities. Yet under the influence of an external force these random moving electrons are found to drift, at a very low speed compared to the chaotic movement, in the direction of the force. We refer to this transport of charge as electric current. Therefore all charge in motion constitutes an electric current. If we desire to measure the current we choose a cross section of the conductor and measure the continuous flow of electrical chage across this boundary in any given period of time. Thus an electric current is defined as the time rate of motion electric charge across any cross sectional boundary. The flow of electricity along a conductor may be measured by the number of coulombs of electricity which flow across a given cross section of the conductor in one second. The coulomb is a certain definite quantity of electricity and it contains a certain definite number of electrons. Experimental evidence shows that each electron contains 1.602×10^{-19} coulombs of charge. There are, therefore, about 6.24×10^{18} electrons in a coulomb. Since in most electric circuits the motion of electrical charges involve terrifically large numbers of electrons, the coulomb was chosen in order to use smaller numbers in electrical calculations. Therefore the electric current is measured in coulombs per second, not electrons per second. Even the coulomb per second is looked upon as too cumbersome. Hence, the rate of flow of one coulomb per second is defined as one ampere. Benjamin Franklin chose to define positive sense of current as the direction of positive charge motion. The positive direction of current is, therefore, considered, to be opposite to the direction in which electrons move. This convention, of considering the positive direction of current as the direction in which protons would move if they were free to do so, remains with us to lay. If all currents were caused only by electronic conduction, we would consider a change in the convention desirable but we have seen that current may also be caused by the flow of positively charged ions. On a circuit diagram, which is really a symbolic abstraction of a practical electrical system, the current direction is indicated by means of arrowhead on the interconnecting wires which carry the current or by means of arrows drawn besides the wires. In the figure 1 below the direction of the current flow is counter clockwise, as shown by the arrows and labelled I. The figure demonstrates an electrical system comprising a battery, an electrolytic solution, a vacuum tube and the interconnecting wires. In the electrolytic solution the current consists of positive ions whose motion is in the same direction as I and negative ions moving in the opposite direction. In the wires and in the vacuum tube the current consists of electrons moving in the direction opposite to I. Mathematically, the current is related to the charge and time by the expression from

the calculus $i = \dfrac{dq}{dt}$ amperes   (Equation 1.4)

Where i is the current in amperes
     q is the Charge in coulombs
     t is the time in seconds

By international agreement the symbol I is used for current. The ampere is therefore the common unit for measuring the rate of flow of electricity. When we say that 100 watt, 110 volt incandescent lamp takes a current of approximately one ampere we mean that as long as the lamp is glowing with its normal intensity a current of about one ampere is flowing through it continuously. Electrical currents are generally measured by electromechanical devices known as ammeters which may be inserted into circuits with a minimum disturbance to the circuit. The current is said to be continuous if it does not change in magni-

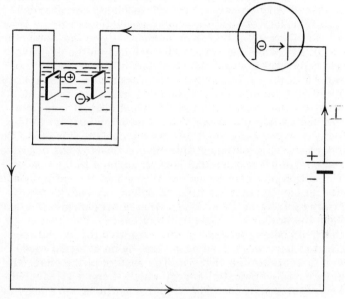

Figure 1

tude with time. A direct current is one which flows in one direction only and never reverses. A direct current may change in magnitude with time, even though it does not change its direction. A pulsating current is a direct current changing in magnitude with time. An alternating current is one which alternates periodically; that is, a periodic current in which changes in direction and magnitude with time, unless otherwise specified, are repeated regularly. An alternating current is said to have the same wave form in every period. In dealing with direct currents in this text we shall confine ourselves to continuous direct currents.

1.5 Electromotive Force and Potential Difference

Whenever an electric current flows it is accompanied by an interchange of energy. We know that when a neutral, uncharged, conductor is placed close to a charged body there occurs a redistribution of the free electrons in the conductor until equilibrium is established. If charged body carries a positive charge the free electrons in the neutral conductor are attracted and will concentrate on that portion of the conductor

closest to the charged body. Now suppose we have a neutral conductor placed very closely to, but not touching, two charged
bodies, one negatively and the other positively charged, a flow
of the free electrons takes place momentarily in the conductor.
This occurs because the excess of electrons on the negatively
charged body tend to repel the electrons of the conductor near
that end while the excess protons on the positive charged body
tend to attract the electrons of the conductor near that end. If
now we allow the conductor to make actual contact with both
charged bodies, the electrons of the conductor nearest to the
positively charged body move across the contact surface to the
positive charges; simultaneously, the electrons of the negatively
charged body move across that contact surface to the conductor. The action is a continuous one throughout the entire conductor until an equilibrium condition of balance of charge has
occurred resulting in the neutralization of all charges. We may
look upon the action as a conversion of energy. The two origin-
ally charged bodies, because of the attractive forces between
them, had to have work done upon them by an external force
in order to separate the charges. The system of the charged
bodies then gained potential energy which it could release under the proper conditions. The proper conditions were provided
when the conductor was placed in contact with the two charged
bodies, resulting in the release of the stored energy and its
conversion into heat.

If we desire a continuous transfer of charge in the conductor, and we do in order to maintain a continuous current flow,
we see that it becomes necessary to obtain a device capable of
maintaining a pair of bodies in oppositely charged state even
when they are externally connected by a conductor. This device
must be capable of maintaining the condition of the charged
bodies by motion of properly equivalent charge through the device itself and this separation of charges can be accomplished
only by the conversion of energy from some other form to electrical energy within the activating device. Alessandro Volta, an
Italian physicist, developed such an activating device, capable
of maintaining a continuous flow of current, known as the Voltaic
cell or battery. We shall refer to the activating device as a
source of electrical energy. We will further restrict ourselves
to the conventional positive charge. Thus, conventional positive
charges receive an increase in their electric potential ability
to do work when they receive energy from an electric energy
source. The electric charges actually possess an electric potential rise. By convention we assign the electric potential rise
to the source and consider the source as the cause of the elec-

tric action. It is customary call this potential rise an electromotive force. We generally use the abbreviation emf to
describe an electromotive force. Consider the electical circuit
of Figure 2. It is made up of an electrical energy source and an
external device, known as a load, which is capable of converting
electrical energy into some other form, for example such as
heat. The source and the load are joined together by perfect conductors. The electrical charges in flowing through the circuit
external to the source release the potential energy they had
previously acquired. A little reflection will show that for a completely closed circuit the drop in electric potential through the
load and the interconnecting conductors must be balanced by
the rise in the electrical potential through the source.

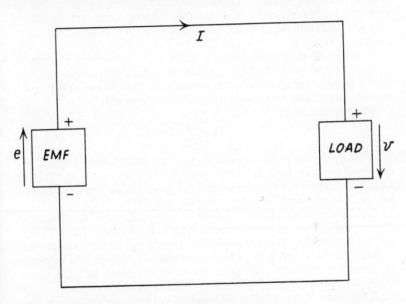

Figure 2

To recapitulate, whenever electric charges are separated there
must be present some external force which is capable of converting energy from a non electrical into electrical form. Thus
there must be some motive power which tends to produce a current. Such an action is present in all electric generators, chemical cells and thermo-couples.

Quantitatively, we define an electromotive force as follows;
The Emf of an electrical energy source is said to possess one

Volt when one Coulomb of charge is transported through the source in the direction of the rise in potential and is accompanied by the conversion of one Joule of energy to the electric form.

In mathematical form,

$$e = \frac{dw}{dq} \quad \text{(volts)} \qquad \text{Eq 1.51}$$

where dw is the energy in Joules which is given by the non electrical system to a very small charged particle dq in coulombs, and e is the emf in volts.

Similarly we define the electric potential drop which is commonly referred to as the voltage drop of an electrical load as

$$v = \frac{dw}{dq} \quad \text{(volts)} \qquad \text{Eq 1.52}$$

where dw is the energy in Joules which is given by the electrical system, to be converted to some other form, to a very small charged particle dq in coulombs, and v is the voltage drop in volts.

Thus in an electrical circuit, when one coulomb of electricity gains or loses one joule of energy in being transported between two points in the electrical circuit, the potential difference between the two points is said to be one volt.

Mathematically

$$V_{12} = \frac{W}{Q} \quad \text{(volts)} \qquad \text{Eq 1.53}$$

where V_{12} is the potential difference between points 1 and 2 in the electrical circuit; Q is the quantity of electricity in coulombs which is transferred from point 1 to 2; and W is the energy in joules lost or gained by Q during the transfer.

Voltage measurements are generally made by electrical instruments known as volt-meters. These instruments are connected across the two points in the electrical circuit where the voltage measurement is desired. By proper design and selection of these instruments it is possible to insert them into the electrical system with negligible disturbance of the system.

If we know the potential difference between two points in an electrical circuit we can calculate the amount of energy gained or lost by an electron which moves from one point to the other, by multiplying the electronic charge, Q_e, by the potential difference.

Thus

$$W = V_{12}\, Qe \quad \text{(Joules)} \qquad \text{Eq 1.54}$$

1.6 Power Relationship

We will now consider one of the most useful relationships in electrical technology, namely electrical power. We have seen that the quantitative relation for the electromotive force is given by

$$e = \frac{dw}{dq} \text{ (volts)} \qquad \text{(Eq. 1.51)}$$

and the quantitative relationship for the current is expressed by

$$i = \frac{dq}{dt} \text{ (amperes)} \qquad \text{Eq. 1.4}$$

Thus for the electrical energy source the product of the Emf and the current yields the rate of transfer of electrical energy into the system:

$$ei = \frac{dw}{dq} \times \frac{dq}{dt} = \frac{dw}{dt} \text{ Joules/sec. or watts}$$

Therefore the power input to the electrical circuit is given by the rate of transfer of the electrical energy. Mathematically we say

$$p = ei \text{ (watts)} \qquad \text{Eq. 1.61}$$

where p is the electrical power input in watts, e is the emf in volts and i is the electric current in amperes.

At the electrical load, the electrical power output of the circuit is found by forming the product of the voltage drop across the load and the current flowing through the load. Quantitatively,

$$p = vi \qquad \text{(watts)} \qquad \text{Eq. 1.62}$$
$$= \frac{dw}{dq} \times \frac{dq}{dt} \text{ (watts)}$$
$$= \frac{dw}{dt} \text{ (watts)} \qquad \text{Eq. 1.62 a}$$

where p is the electric power output in watts
     v is the voltage drop in volts
     i is the electric current in amperes.

Equations 1.61 and 1.62 are very general equations and represent instantaneous relationships. They are correct for all conditions of voltages and currents either constant or varying. We shall have occasion to define many types of power as well as voltages and currents for electrical circuits in which voltages and currents vary with time. But we shall find that all these power definitions are based upon the fundamental laws of instantaneous power as expressed by equations 1.61 and 1.62.

1.7 Polarity Notations

A charged body gains energy in moving from one point to another in an electrical circuit as in Figure 3 where point 2 is at a higher potential thant point 1. If on the other hand the charge losses energy in the motion, point 1 is said to be at a higher potential than 2.

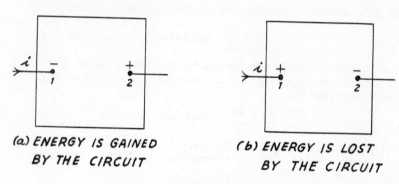

(a) ENERGY IS GAINED
   BY THE CIRCUIT

(b) ENERGY IS LOST
   BY THE CIRCUIT

Figure 3

It is customary to designate the point of higher potential, on a diagram representing the electrical circuit, by a plus sign and the point of lower potential by a minus sign. The plus and minus signs are known as polarity marks. A second method of designating polarity is also commonly used. Double subscripts are used to indicate the potential rise from point 1 to point 2 as V_{12}. If point 2 is actually at a higher potential than point 1, as is the case in Fig. 3 (a), where energy is being gained by the charge in its motion from 1 to 2, than the numerical value is considered positive. In Figure 3a current flows from point 1 to point 2 so that the potential rise from 1 to 2, V_{12}, is positive. For the same circuit the potential rise from point 2 to point 1, V_{21}, is negative. For the circuit of Fig. 3 (b) where the charge losses energy in its motion from point 1 to point 2, the potential rise, V_{12}, is negative and the potential rise from point 2 to point 1, V_{21}, is positive.

It is sometimes more convenient to use potential drops. By convention the potential drop, or voltage drop, between two points such as 1 and 2 in an electrical circuit is designated as V_{12}. By definition the voltage drop, V_{12}, is positive if the actual potential of point 1 is higher relative to point 2. Therefore in Fig. 3a, V_{12}, the voltage drop is negative where as the voltage drop V_{21} is positive. It is customary in the solution of actual problems to use either all potential rises or all potential drops. Except under very special conditions, the use of both in the same problem will result in errors and confusion. It is therefore, a good idea to specify, at the beginning of each problem, whether potential rises or voltage drops will be used.

We will make one additional assumption in this text, by designating the electromotive (Emf) by the symbol E and the potential difference by the symbol V. The potential difference has been defined in terms of energy received in an electric circuit. The electromotive force, E, has been defined in terms of energy given up in a non electrical system.

Devices in which electric power is being generated will also accept electrical energy, that is, they may be operated backwards by sending current through them from their positive terminal to their negative terminal by more potent sources. Electric generators and batteries are some common examples of electrical sources which under proper conditions can become electrical loads. Such devices are referred to as reversible sources. Sometimes, reversible device, containing an Emf, receive power from the electrical system. The Emf is then designated as a "back Emf, to indicate that the power flow is from the electrical system to the non electrical system. An electrical energy source, containing an Emf, does not require an electrical current for its existence. Hence, the value of the Emf is independent of the current and one of the outstanding characteristics of an Emf is that it exists even when the current is equal to zero. Thus, we speak of an Emf as the open circuit voltage of the electrical energy source. By open circuit volvoltage we mean the voltage that exists when there is no current flow.

We may therefore define any device, regardless of the physical process involved, as an electrical energy source when current is directed externally away from its positive terminal and as an electrical load, when current is directed externally toward its positive terminal.

CHAPTER 2—FUNDAMENTAL LAWS OF
ELECTRICAL CIRCUITS

2.1 Ohms Law

The steady flow of the free electrons in a conductor is
hindered by the atomic structure of the conductor. The steady
progress of the free electrons may also be hindered in their motion by collisions with stationary atoms and electrons. These
collisions generally result in a loss of energy, appearing as
heat energy which raises the temperature of the conductor. We,
therefore, generally associate with the flow of current through
a conductor a continuous conversion of electrical energy into
heat energy. We look upon the action that takes place which
hinders the flow of current in a conductor as being very similar
friction. Many devices, such as electrical heaters and broilers
operate on this principle. In many other devices, such as electric motors, the production of heat is undesirable and must be
minimized for the optimum operation of the electrical device.
We reserve a special name for a device in which the flow of
electric current produces only heat. Electric power is always
dissipated in this type of device, which we term a resistor.
Hence the current through a resistor must always flow from a
terminal of higher potential to a terminal of lower potential.
That is, the resistance to the flow of current through a conductor requires a potential difference between the terminals of any
practical conductor. If the current in a resistor is reversed
the potential difference must reverse. It follows, therefore, that
if the current through a particular resistor is zero the potential
difference must be zero. Dr. George Simon Ohm studied the relationship between the potential difference and the current experimentally and in 1826 published the results of his investiga-
tion. Dr. Ohm discovered that for a particular metallic conductor, held at a constant temperature, the relationship between the
potential difference across the terminals of the conductor and
the current through the conductor is a linear one. Thus, if the
potential difference is doubled, the current is also doubled. When
a curve is drawn, making a graph of potential difference against

current, a straight line, passing through the origin, is obtained. The slope of this line is constant and represents the ratio of the potential difference across the conductor to the current in the conductor. This ratio is called the resistance of the conductor. The symbol R is used to designate resistance.

Mathematically this relationship is expressed as:

$$\frac{V}{I} = R \text{ (ohms), a constant} \qquad \text{Eq. 2.1}$$

Where V is measured in volts, I is measured in amperes and R is measured in ohms. The unit of resistance, the ohm, is named in honor of Dr. Ohm and the relationship is known as Ohm's Law. Ohm's law for a metalic conductor at a constant temperature may be stated as:

$$R = V/_R \text{ ohms} \qquad \text{Eq. 2.1}$$

or $$\qquad I = V/_R \text{ amperes} \qquad \text{Eq. 2.1a}$$

or $$\qquad V = IR \text{ volts} \qquad \text{Eq. 2.1b}$$

The ratio of the current through a metallic conductor to the potential difference across the terminal of the conductor, at a constant temperature is defined as the conductance or the amperes per volt. The letter G is used for conductance and is also known as the reciprocal ohms or mhos. Mathematically, we say that

$$G = \frac{1}{R} \text{ (mhos)} \qquad \text{Eq. 2.2}$$

and Ohm's law may also be expressed as:

$$G = \frac{I}{V} \text{ (mhos)} \qquad \text{Eq. 2.2a}$$

$$V = \frac{I}{G} \text{ (volts)} \qquad \text{Eq. 2.2b}$$

$$I = VG \text{ (amperes)} \qquad \text{Eq. 2.2c}$$

Dr. Ohms also investigated the relationship between the temperature of metallic conductors and their resistance. He found that for most metals and their most common alloys the resistance increased when the temperature was raised. This type of characteristic lends itself very advantageously to the design and operation of the incandescent lamp for lighting. As the current is increased in a metallic conductor, such as a lamp filament, the energy which is being converted into heat is increased, which in turn increases the temperature. An increase in temperature tends to increase the resistance. This results in a potential difference which does not increase in proportion to the current.

Here we find that the straight line relationship between potential difference and current no longer exists.

Since the material of a conductor will tend to influence the ease or difficulty with which the free electrons may be moved within the conductor, the different conductors tend to have different resistances to the flow of current. But for most metallic conductors, at constant temperature, Ohm's law is valid. For many non-metallic materials the ratio of the potential difference to the current is not a straight line. For example, in Figure 4 (page 42) we have plotted typical volt-ampere relationships for a metallic conductor, for a lamp filament conductor and for a semi-conductor. It can be seen that the semi-conductor volt-ampere characteristic does not follow a straight line relationship. The curvature here however is not due to temperature changes but rather to the manner in which electrons behave when only a limited number are free electrons.

A resistor whose volt-ampere curve is not a straight line is called a non linear resistor. Non linear resistors are becoming extremely important in electrical technology. We have but to mention such non metallic substances as thermistors, vacuum tubes, crystal rectifiers and transistors to recognize the importance they play in electrical control and communication.

2.2 Joule's Law

James Prescott Joule, an English physicist, was one of the pioneers who helped establish the idea of the conservation of energy. His experimental researches revealed that when a current exists in a metallic conductor the heat developed in a given time is equal to the resistance of the conductor multiplied by the square of the current. Thus, the rate at which heat is evolved in the conductor is proportional to the square of the current flowing through the conductor. Moreover, this energy transfer, from the electrical form to heat is not reversible. By definition, power is the rate of doing work, i.e., the total amount of work done, or energy converted in given time is given by the product of the power and the time. If the power is steady we may say

$$W = Pt \quad \text{(Joules)} \qquad \text{Eq. 2.3a}$$

W is the energy converted during the time t, in seconds, P is the power in watts.
A 100 kilowatt generator operating at full load for 10 hours will deliver 1000 kilowatt-hours of electrical energy. The units of energy are usually formed by combining the units of power with the units of time. That is, the watt-second is a watt of power

maintained for one second. When the power is not steady, we must resort to the calculus to obtain the energy. Mathematically, by integration we obtain,

$$W = \int_{T_1}^{T_2} P \, dt \ \text{(Joules)} \qquad \text{Eq. 2.3b}$$

where W is the energy in Joules converted during the time interval $T_2 - T_1$, in seconds and P is the power in watts.

Joules' law can be stated mathematically as

$$P = I^2 R \ \text{(watts)} \qquad \text{Eq. 2.4a}$$

where P represents the rate at which heat is developed in a metallic conductor, in watts; R is the resistance of the conductor in ohms and I is the current flowing through the conductor in amperes. This law is valid when the current is steady or is varying If the current does change in magnitude with time, the law will still hold provided it changes so slowly that current density over the entire cross section of the conductor remains constant.

2.3 Power Converted in a Resistor

From Ohm's law we see that the potential which will cause a current of I amperes to flow in a resistance of R ohms can be expressed as, the product, IR, from the definition of potential difference. As derived in equation 1.61, the net power which is being converted into heat in a resistor is expressed as

$$P = VI \ \text{(watts)} \qquad \text{Eq. 1.61}$$
and since
$$V = IR \ \text{(volts)} \qquad \text{Eq. 2.16}$$
therefore
$$P = I^2 R \ \text{(watts)} \qquad \text{Eq. 2.4a}$$

Since it is also possible to state Ohm's law as

$$I = {}^V\!/_R \ \text{(amperes)} \qquad \text{Eq. 2.1a}$$

there
$$P = \frac{V^2}{R} \ \text{(watts)} \qquad \text{Eq. 2.4b}$$

The power relationship may therefore be stated mathematically in forms:

$$P = I^2 R \ \text{(watts)} \qquad \text{Eq. 2.4a}$$
and
$$P = \frac{V^2}{R} \ \text{(watts)} \qquad \text{Eq. 2.4b}$$

or in the most general form:

$$P = VI \ \text{(watts)} \qquad \text{Eq. 2.4c}$$

This form is most applicable when a current of I amperes flows due to a potential difference of V volts, then the power delivered or received will be given by P watts.

The two other forms are applicable only when V is the potential difference across a resistance of R ohms due to current of I amperes. Then the power P is given by I^2R and $\frac{V^2}{R}$.

We have already seen that the conversion of electrical energy into heat energy has many useful applications, such as illumination, heating, etc. However, it is important to keep in mind, that whenever we desire to conduct electric current from one place to another we must use a metallic conductor such as copper wires. This means that we will always have power loss in the form of heat loss in conductors. Since, this heat loss is an I^2R loss we deliberately choose conducting wire such as copper or aluminum which have a very small value of resistance and thus reduce the heat losses in the conductors. The I^2R loss in conductors is all converted into heat in the same way in which all the mechanically energy used by a machine to overcome friction is changed into heat.

Equation 2.3b represents the mathematical expression of energy

$$W = \int_{T_1}^{T_2} p\,dt \text{ (Joules)}$$

and when expressed as

$$W = \int_{T_1}^{T_2} i^2R\,dt \text{ (Joules)} \qquad \text{Eq. 2.5}$$

will always represent heat. It can therefore be readily expressed as heat units in calories. For a continuous direct current the expression becomes

$$W = I^2R\,t \text{ (Joules)} \qquad \text{Eq. 2.6a}$$

Since it is known that 1 Joule equals 0.24 calorie:

$$H = 0.24\,I^2R\,t \text{ (calories)} \qquad \text{Eq. 2.6b}$$

where H represents the heat generated in calories, I the current in amperes, R the resistance in ohms and t the time in seconds.

2.4 Power Converted in an EMF.

An electrical circuit comprises a source of electrical energy and an electrical load, joined by the necessary inter connecting wires. The energy which is converted into heat in the interconnecting wires represents an irreversible flow while the energy flow which is converted in an Emf is always reversible.

Therefore, power can be made to flow from a non electrical form into an electrical form or the process may be reversed. We will consider electrical power as being generated whenever non electrical energy is being converted into electrical energy. Hence we will consider electrical power as being consumed whenever electrical energy is being converted into non electrical form. At a generating station, for exampe, mechanical energy is being transformed into electrical energy. This exchange of energy may take the form of the conversion of the heat of high pressure steam into electrical form. We may find the time rate at which this conversion occurs by forming the product of the current and the Emf. In mathematical form

$$P = EI \text{ (watts)} \qquad\qquad \text{Eq. 2.7}$$

where P represents the power in watts, E is the electromotive force (Emf) in volts and I is the current in amperes.

If we make use of the polarity notation which we defined in Chapter 1, we may assign a physical meaning to the algebraic sign of P. If E represents a voltage rise in the direction of the current flow, then when P is positive it will mean that power is being generated; a negative P, would then mean that power is being used up. We may, however, also consider E as a voltage drop in the direction of current flow. This, of course, means that a positive P will be considered as power being consumed and a negative P as power which is generated.

2.5 Kirchhoff's Laws

The flow of electric currents in a circuit or network of conductors was studied by a German scientist, Gustav Kirchhoff. He formulated two very general laws concerning the flow of electric currents. These laws are based upon the two fundamental principles, namely, the law of conservation of matter and the law of conservation of energy. In fact, we may say, that Kirchhoff's laws represent the application of these principles to the flow of electric currents in electrical circuits.

2.51 Kirchhoff's Current Law

Many elements forming an electrical circuit may have a common connecting point. A common terminal is generally referred to as a junction point. Of course.the mathematical point really represents any small volume of conducting path in any physical circuit. There is sufficient experimental evidence to show that the amount of direct current leaving a junction point in a circuit is equal to the amount entering that junction point.

We recognize here a re-statement of the law of conservation of matter. Since current is the time rate of motion of charge in a circuit and here we are dealing with electron flow, there are no electrons lost in this process. At any junction point in a circuit just as many electrons arrive every second as are being transported away, when charge does accumulate at the junction. The junction point, of course, is assumed to be very small and incapable of storing any electrical charge. In any direct current circuit, we say that electricity will not accumulate at a point. Hence, Krichhoff's current law can be visualized from Figure 4a

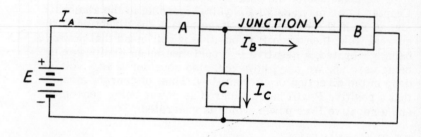

Figure 4a

which represents an Emf delivering electrical power to three electrical loads which have a common junction point at Y. Here the current I_A flowing to the junction Y must be exactly equal to the sum of currents I_B and I_C which are flowing away from the junction Y. We may express this relation by the equation

$$I_A = I_B + I_C \text{ (amperes)} \qquad \text{Eq. 2.8a}$$

Kirchhoff's current relationship is quite general hence it will also be true for currents which are changing in magnitude and direction with time as well as for steady continuous currents.

Hence if we consider instantaneous currents as in figure 4b, we will express Kirchhoff's current law as

$$i_{12} = i_{23} + i_{24}$$

Eq. 2.8b

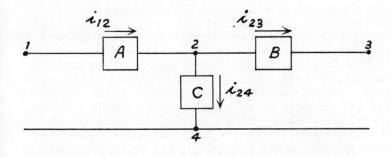

Figure 4b

Figure 4b represents 3 circuit elements having a common junction at 2. The currents are considered to be flowing instantaneously in the directions indicated through the circuit elements.

We could also have written the equation 2.8b as

$$i_{12} - i_{23} - i_{24} = 0$$

Eq. 2.8c

We may arbitrarily call those currents which are directed toward a junction as positive and those leaving as negative. By a negative current we mean, of course, a positive current which is leaving a junction. Hence, Kirchhoff's current law may be stated as follows:

The algebraic summation of all the currents entering a junction point is zero. Mathematically we may express this as

$$\Sigma I = 0$$

Eq. 2.8d

2.52 Kirchhoff's Voltage Law

Kirchhoff's voltage law is simply an expression for the principle of conservation of energy. If an electric potential difference exists between any two points in an electrical circuit it will exert the same influence on all paths which may be constructed to connect the two points. In other words, the electric potential difference between any two points is the same regardless of the path along which it is measured. This statement follows from the fact that the potential difference is really a measure of the work done in moving an electrical charge between the two points. Consider the circuit which is shown in Figure 5. It comprises four two terminal circuit elements; each one of the elements may consist of various circuit components such as Emf's and simple current carrying resistors singly or in any combination. The four elements are connected to form a complete circuit. We will generalize by using the symbol v to represent an electric potential drop and in accordance with the polarity notation we chose to use in chapter 1, V_{12}, represents point 2 at a lower potential than point 1 regardless of the cause which may be an Emf, a resistance through which current is flowing or a net combination of the two. Kirchhoff's voltage law may then be summarized as follows:

The algebraic summation of the voltage drops around a closed path is zero. In Figure 5, we start our closed path at point 2, the Kirchhoff voltage law becomes, traversing our path in a clockwise direction:

$$- v_{12} + v_{34} + v_{56} + v_{78} = 0 \qquad \text{Eq. 2.9a}$$

Here we have designated, quite arbitrarily, voltage drops as positive and voltage rises as negative. Of course, if we had traversed our path in a counter clockwise direction we would have given positive signs to the voltage rises and negative signs to the voltage drops as follows:

$$v_{12} - v_{78} - v_{56} - v_{34} = 0 \qquad \text{Eq. 2.9b}$$

This is an equally valid equation since the summation is equal to zero.

Kirchhoff's voltage law is, of course, generally true for time varying as well as steady voltages. For time varying voltages, we will consider v_{12}, as an instantaneous voltage drop. Kirchhoff's voltage law can also be visualized from the electrical circuit shown in Fig. 6

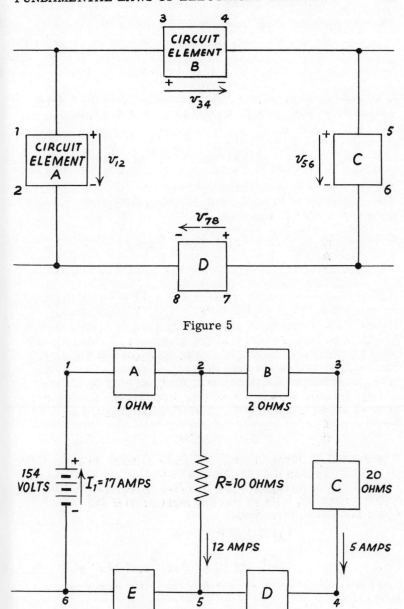

Figure 5

Figure 6

Thus the potential drop from 2 to 5 is given by Ohm's law as

$$V_{25} = I_{25} R_{25} \text{ (volts)} \qquad \text{Eq. 2.1b}$$

hence

$$V_{25} = 12 \times 10 = 120 \text{ (volts)}$$

If we had found the voltage drop by choosing the path 2-3-4-5 instead of the path directly from 2 to 5, we would have

$$
\begin{aligned}
V_{25} &= V_{23} + V_{34} + V_{45} \text{ (volts)} \\
&= I_{23} R_{23} + I_{34} R_{34} + I_{45} R_{45} \\
&= 5 \times 10 + 5 \times 20 + 5 \times 2 \\
&= 120 \text{ volts}
\end{aligned}
$$

If we now proceed to choose still another path such as 2-1-6-5-, the equation for V_{25} becomes

$$V_{25} = V_{21} + V_{16} + V_{65}$$

$$= I_{21} R_{21} + V_{16} + I_{65} R_{65}$$

$$= -17 \times 1 + 154 - 17 \times 1$$
$$= 120 \text{ volts}$$

We note from these examples that Ohm's law is contained in the Kirchhoff voltage law. We consider a resistor, R, whose symbolic notation as a circuit element is designated in Figure 6. The resistor will be assumed to carry a current, $I = 12$ amperes, in the direction shown. In accordande with Ohm's law and our polarity notation the voltage drop through the resistor is IR as shown in the figure, and which we designated as:

$$V_{25} = I_{25} R_{25}$$

Suppose that we form a closed path by starting at terminal 2 and go directly through the resistance to 5 tracing in clockwise direction and return through the path 5-4-3-2 to terminal 2. Kirchhoff's voltage law tells us that this summation of the voltage drops is equal to zero. Thus

$$V_{25} = I_{25} R_{25}$$

and

$$V_{25} = I_{23} R_{23} + I_{34} R_{34} + I_{45} R_{45}$$

hence the voltage drops around the completely closed path is seen to be:

$$I_{25} R_{25} + I_{54} R_{54} + I_{43} R_{43} + I_{32} R_{32} = 0$$

or

$$12 \times 10 - 5 \times 2 - 5 \times 20 - 5 \times 2 \qquad = 0$$

CHAPTER 3—APPLICATION OF KIRCHHOFF'S LAWS
TO DIRECT CURRENT CIRCUITS

We make use of Kirchhoff's laws in the solution of electrical circuit problems. Any direct current electrical problem regardless of the many possible complicated connections can be solved by means of the Kirchhoff's relationships, if they are properly applied. When these relationships are properly interpreted they are used to solve alternating current circuit problems too.

We have previously stated the main function of electric circuit theory is to provide a method of determining voltages and currents and their associated quantities such as power and energy, throughout the electric circuit. The electric circuit, itself, is a physical arrangement of the generators of electrical energy, the apparatus for the utilization of this energy, which we have termed as electrical loads and their physically interconnecting wires. The solution of such a physical system has been simplified by introducing a symbolic notation for these electrical components and drawing a symbolic representation in a plane, such as a page of this book. To relate these symbols to the physical nature of the circuit they have been selected to represent concentration of energy storage, either electric or magnetic, energy dissipation and energy of the reversible type. Symbols that are so defined are known as circuit elements. The elements that represent electrical sources that are capable of sustained energy conversion that is reversible are given the name of active elements; where as those elements which represent simply storage or dissipation of energy are known as passive elements. Thus, electromotive forces (emf's) are example of active elements while resistors, capacitors and inductors represent passive elements. Passive elements are also regarded as acquiring no energy for storage or dissipation except from direct or indirect connection to an active element. Passive circuit elements can not acquire current or voltage except by such means.

In applying the Kirchhoff's laws to such electrical circuits a great deal of labor in computation can be saved when a systematic approach is used. We may specify certain general rules for the methods used in the solution of electrical circuits which will

help us to attain both speed and accruacy. First, make a good
circuit chagram. Indicate on this diagram every constant that is
known, i.e., resistance values, potentials and currents. Indicate
potentials by + and - signs, and the direction of currents by arrows. Assign letters to the unknown quantities. In accordance
with the symbols we have already defined, call currents I, potentials V, emf's E, and resistance R. Use double subscripts when
ever possible. Assume the direction for each unknown current
and mark this with an arrow. Similarly put + and - marks on
each V and E to indicate the polarity which you have assumed.
Use the minimum number of unknowns, just a sufficient number
to be able to write just enough equations to solve for each circuit. Now you are ready to apply Kirchhoff's laws to parts of
the complete circuit to obtain equations between the unknowns.
Write only as many equations as there are unknowns.

Since we will use both Kirchhoff's current and voltage relations we may list some rules for the number of independent
equations which can be written. A junction point is also known
as a node and we recall that it represents any point in the circuit where two or more currents join to form a third. If we
write all the possible current equations in an electrical circuit
we find that the number of independent equations is one less
than the number of nodes.

We also define a branch as any section of a circuit which
joins two nodes directly, without passing through a third node.
The number of independent voltage equations that can be written
may be obtained from the number of branches and the number
of nodes in an electrical circuit. In using the voltage equations
we find that the number of independent voltage equations will be
given by the number of branches.less the number of independent
node equations. In the application of the Kirchhoff voltage equations the summation of the voltage drops may of course be about
any closed path or loop. Choose the obvious and the simplest
loops, making sure, of course, that the unknown voltages and currents are used. The simultaneous solution of the algebraic equations will yield the unknown quantities. One of the best systematic methods for the solution of simultaneous algebraic equations
makes use of determinants. We will use the method of determinants in this text.

3.1 Series and Parallel Combinations of Resistances

A direct current circuit which contains nothing but resistors
is generally called a two terminal passive network. Many complicated resistance networks can be reduced to a simple two

terminal passive network with a single resistance between the
terminals. Consider the circuit shown in Figure 7a. Here we see

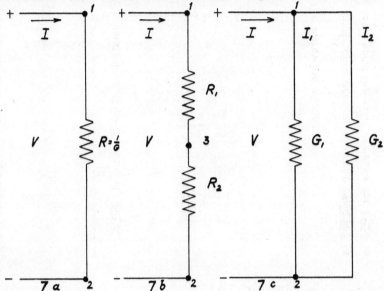

the simplest form of the two terminal resistance network. From
Ohm's law we obtain:

$$V = RI$$

and $$I = GV$$

Figure 7b represents two resistance elements in series while
Figure 7c represents two conductance elements which are connected in parallel. In Figure 7b it is obvious that the same current, I, must flow through both resistance elements. This can
be obviously seen from the application of Kirchhoff's current
law at node 3. Now applying the voltage law to the entire circuit
we obtain:

$$V - R_1 I - R_2 I = 0$$

or $$V = I(R_1 + R_2)$$

Whence, we obtain

$$R = R_1 + R_2 \text{ (ohms)}$$

Therefore the circuit of Figure 7b will behave exactly as the
one of Figure 7a if R is equal to the sum of R_1 and R_2. If there
had been any number of resistors in series between terminals

1 and 2, the resultant series resistance between the two terminals, which we label as R_{se}, would be given by:

$$R_{se} = R_1 + R_2 + \ldots R_n \qquad \text{Eq. 3.1}$$

Since the conductance, G mhos, represents the reciprocal of the resistance then

$$G = {}^1/R$$

whence

$$R = R_1 + R_2 = \frac{1}{G_1} + \frac{1}{G_2}$$

$$= \frac{G_1 + G_2}{G_1 G_2}$$

or

$$G = \frac{G_1 G_2}{G_1 + G_2} \quad \text{(mhos)}$$

Hence for any number of conductances in a series connection, the equivalent series conductance, G_{se}, will be given by the quite complicated euation:

$$G_{se} = \frac{1}{\dfrac{1}{G_1} + \dfrac{1}{G_2} + \ldots + \dfrac{1}{G_n}} \quad \text{(mhos)} \qquad \text{Eq. 3.2}$$

Due to this complicated expression as compared to the simple one for the series equivalent resistance, the series equivalent conductance is seldom used. We now consider the circuit of Figure 7c which consists of two conductances connected in parallel. It can be readily seen from the application of Kirchhoff's voltage law that the same voltage exists across both conductors. Applying Kirchhoff's current law at node 1 we get:

$$I - I_1 - I_2 = 0$$

or

$$I = I_1 + I_2$$

but

$$I_1 = G_1 V$$

and

$$I_2 = G_2 V$$

whence

$$I = (G_1 + G_2) V \text{ (amperes)}$$

but in the circuit of Figure 7a

$$I = GV \text{ (amperes)}$$

therefore if

$$G = G_1 + G_2 \text{ (mhos)}$$

the two circuits will behave in an identical manner. Therefore the equivalent parallel conductance, G_{pe}, will be given by

$$G_{pe} = G_1 + G_2 \text{ mhos}$$

Therefore for any number of conductances in parallel, G_1, G_2, G_n, the equivalent parallel conductance will be given by

$$G_{pe} = G_1 + G_2 + ... + G_n \text{ (mohs)} \quad \text{Eq. 3.3}$$

Since the resistance is the reciprocal of the conductance

$$G = G_1 + G_2 = \frac{1}{R}$$

may be written as: $\quad G = \dfrac{1}{R_1} + \dfrac{1}{R_2}$

$$= \frac{R_1 + R_2}{R_1 + R_2}$$

whence

$$R = \frac{R_1 \ R_2}{R_1 + R_2} \quad \text{(ohms)}$$

and for any number of resistances in parallel connection, the equivalent parallel resistance is given by:

$$R_{pe} = \frac{1}{\dfrac{1}{R_1} + \dfrac{1}{R_2} + \cdots \cdots + \dfrac{1}{R_n}} \text{ (ohms)} \qquad \text{Eq. 3.4}$$

Thus any series—parallel combination of resistances can be reduced to a single equivalent resistance by the application of equations 3.1 and 3.4.

3.2 Numerical example 1

A direct current generator delivers power to a set of resistance loads which are connected as indicated in Figure 8. A voltmeter, which does not disturb the system, is found to measure 40 volts across R_1 . G_2 and G_3 are known to be 0.04 and 0.06 mho respectively. It is desired to find the value of the currents I_{12}, I_{34} and I_{56}. It is also desired to find the power which is dissipated in heat in the resistance R_1.

We can apply Kirchhoff's current law of node 2 and obtain the relation:

$$I_{12} - I_{34} - I_{56} = 0$$

We also apply Kirchhoff's voltage law to the closed loop 1-2-7-1 and obtain the relation:

$$100 - V_{12} - V_{27} = 0$$

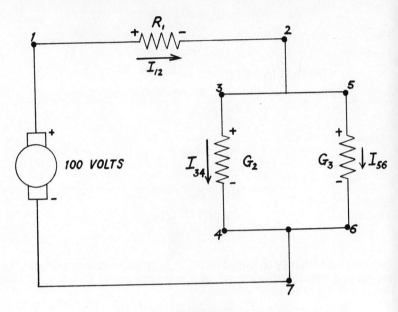

Figure 8

and since V_{12} is known to be 40 volts we have:

$$V_{27} = 60 \text{ volts}$$

and this voltage appears across the two conductances G_2 and G_3 which are connected in parallel. The equivalent conductance is therefore:

$$G_{pe} = G_2 + G_3$$

$$= 0.10 \text{ mhos}$$

and since

$$R = \frac{1}{G} = 10 \text{ ohms}$$

The equivalent resistance for the branch 2-7 is 10 ohms. Hence the current flowing through branch 2-7 can be found from Ohm's law:

$$I_{27} = \frac{V_{27}}{R}$$

$$I_{27} = \frac{60}{10} = 6 \text{ amperes}$$

but from Kirchhoff's current law I_{27} is equal to I_{12} and hence:

$$I_{12} = 6 \text{ amperes}$$

Since V_{12} is 40 volts and I_{12} is 6 amperes from Ohm's law:

$$R_1 = \frac{V_{12}}{I_{12}} = 6.67 \text{ ohms}$$

The power dissipated in heat in the resistance R_1 is found from the relation:

$$P = I_{12}^2 \; R \quad \text{watts}$$
$$= (6)^2 \; (6.67)$$
$$= 240 \quad \text{watts}$$

Since the power is also given by the product of the potential drop across the resistor and the current through the resistor,

$$P = V_{12} \; I_{12} \quad \text{watts}$$
$$P = (40) \; (6)$$
$$P = 240 \quad \text{watts}$$

and checks our initial computation . We will now apply Ohm's law to find the currents I_{34} and I_{56}.

$$I = GV \text{ amperes}$$

hence from inspection of Figure 8 we see that

$$I_{34} = G_2 \; V_{27} \quad \text{amperes}$$
$$= (0.04) \; (60)$$
$$= 2.4 \text{ amperes}$$

and
$$I_{56} = G_3 \; V_{27} \quad \text{amperes}$$
$$= (0.06) \; (60) = 3.6 \text{ amperes}$$

We may now check our answers by means of Kirchhoff's current law applied at node 2:

$$I_{12} - I_{34} - I_{56} = 0$$

or
$$6 - 2.4 - 3.6 = 0$$

3.3 Numerical example 2

Two direct current generators are used in a three wire transmission system to deliver power as indicated in Figure 9. It is desired to find the power dissipated in the 1 ohm and 0.8 ohm loads. It is also desirable to know the value of the current in the center wire and the direction in which it is flowing.

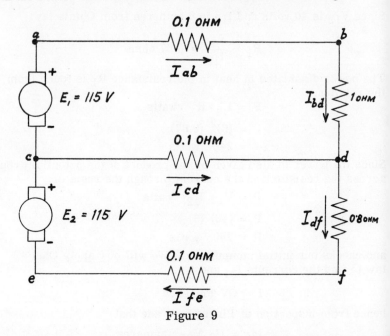

Figure 9

Assume that the branch currents are flowing in the directions indicated in Figure 9. The apply Kirchhoff's voltage law as follows:

For the closed loop a b d c a, using voltage drops:

$$- 115 + V_{ab} + V_{bd} + V_{dc} = 0$$

which upon the application of Ohm's law becomes:

$$- 115 + I_{ab} (0.1) + I_{ab} (1) - I_{cd} (0.1) = 0$$

since it is obvious from Kirchhoff's current law at node b that $I_{ab} = I_{bd}$.

A similar application to the closed loop c d f e c, using voltage drops and applying Kirchhoff's current law at node f to obtain $I_{df} = I_{fe}$ we obtain:

$$- 115 + I_{cd} (0.1) + I_{df} (0.8) + I_{df} (0.1) = 0$$

We now have two independent equations involving 3 unknowns, I_{ab}, I_{cd} and I_{df}, hence we need an additional independent equation for the simultaneous solution of the three unknowns. We can readily obtain such an equation from the application of the Kirchhoff current law at node d:

$$I_{bd} + I_{cd} - I_{df} = 0$$

Hence one of the unknown currents can be eliminated in the two voltage equations and the two equations can be solved simultaneously. Eliminating $I_{df} = I_{ab} + I_{cd}$ in the two voltage equations and simplifying the result we get.

$$(1.1)\ I_{ab} - (0.1)\ I_{cd} = 115$$

$$(0.9)\ I_{ab} + (1.0)\ I_{cd} = 115$$

Solving the two equations by means of determinants we obtain

$$I_{ab} = \frac{\begin{vmatrix} 115 & -0.1 \\ 115 & 1.0 \end{vmatrix}}{\begin{vmatrix} 1.1 & -0.1 \\ 0.9 & 1.0 \end{vmatrix}}$$

$$I_{ab} = \frac{126.5}{1.19} = 106.2 \quad \text{amperes}$$

$$I_{cd} = \frac{\begin{vmatrix} 1.1 & 115 \\ 0.9 & 115 \end{vmatrix}}{\begin{vmatrix} 1.1 & -0.1 \\ 0.9 & 1.0 \end{vmatrix}}$$

$$I_{cd} = \frac{23}{1.19} = 19.3 \quad \text{amperes}$$

Since
$$I_{df} = I_{ab} + I_{cd}$$
$$= 125.5 \quad \text{amperes}$$

We note that the current in the center wire is flowing in the direction from c to d and its magnitude is 19.3 amperes.
The power dissipated in the 1 ohm load is given by

$$P = I_{bd}^2\ R \quad \text{watts}$$

$$= (106.2)^2\ (1)$$

$$= 11,490 \text{ watts or very nearly 11.5 kilowatts}$$

The power dissipated in the 0.8 ohm load will be

$$P = (125.5)^2\ (0.8) \quad \text{watts}$$

$$= 15,600 \quad \text{watts or 15.6 kilowatts}$$

In all of our numerical work we will observe slide rule accuracy. In the computation of practical problems, where data is obtained in the laboratory from actual measurement, slide rule accuracy is generally adequate since our measuring instruments will in general not be any more precise. Ordinary voltages, currents and power can be quickly measured with an accuracy of approximately on half of one per cent. Our normal laboratory instrument will generally be of the two or three per cent accuracy.

3.4 Example 3—The Conversion of Power in a Complete Circuit.

A direct current generator generates an emf of 115 volts and delivers power to a motor as indicated in Figure 10. Since the generator contains wires in its internal structure through which the direct current flows heat will be dissipated in these wires. The generator windings are found to have 1.2 ohms resistance and the lines which are used to connect the generator to the motor have a resistance of 1.5 ohms each. We wish to find how much power is delivered to the motor, and how much mechanical

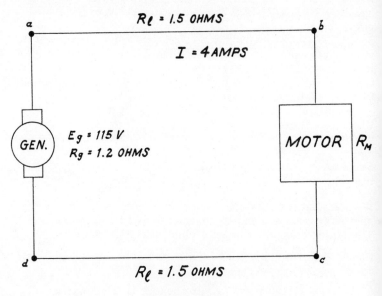

Figure 10

power is developed by the motor if its resistance is 2 ohms.

The current through the motor is measured and found to be 4 amperes. The application of the Kirchhoff voltage law to this circuit, using voltage drops yields:

$$- E_g + IR_g + IR_{ab} + IR_{cd} + E_{bc} + IR_{bc} = 0$$

where E_g is the generated emf;

IR_g is the voltage drop in the wires of the generator,

IR_{ab} and IR_{cd} represent the voltage drop in the lines,

IR_{bc} represents the voltage drop in the wires of the motor; E_{bc} represents the "back emf" of the motor and is represented as a voltage since a motor converts electrical energy into mechanical energy.

We may obtain the power conversion in this circuit if we multiply the equation by the current, I.

$$-IE_g + I^2R_g + I^2(R_{ab} + R_{cd}) + I^2R_{bc} + E_{bc} = 0$$

Here we see that:

IE_g is the power which is converted from mechanical to electrical power in the generator; I^2R_g is the power which is dissipated as heat in the wires of the generator; $I^2(R_{ab} + R_{cd})$ represents the power dissipated in heat in the transmission wires; I^2R_{bc} is the heat dissipated in the wires of the motor and IE_{bc} is the electrical power converted into mechanical power in the motor. The algebraic summation of all of these powers is, as is seen in the equation, equal to zero. This is a very important and fundamental conclusion because it says that in a direct current circuit, when we neglect the transient condition, electric power flows in and out at the same rate and no electric power is stored.

In general, making use of Kirchhoff's current law also, a little reflection will show that in a direct current circuit all the power which is generated is used up, so that the net power is zero.

From an application of the Kirchhoff voltage law we see that the potential across the motor is given by.

$$V_{bc} = E_g - I(R_g + R_{ab} + R_{cd})$$
$$= 115 - 4(1.2 + 3)$$
$$= 98.2 \quad \text{volts}$$

Power delivered to the motor is found from the potential across the motor and the current through the motor, thus:

Power to the motor $= (98.2)(4)$
$$= 392.8 \quad \text{watts}$$

The total power which is generated is given by

$$IE_g = (115)(4)$$
$$= 460 \quad \text{watts}$$

The power consumed in heat by the internal resistance of the generator is

$$I^2R_g = (16)(1.2)$$
$$= 19.2 \quad \text{watts}$$

The power which is delivered to the line is
$$I E_g - I R_g = 440.8 \quad \text{watts}$$
The power dissipated in heat in the transmission lines is of
course, the power delivered to the lines less the power delivered
to the motor:
$$440.8 - 392.8 = 48 \quad \text{watts}$$
It may also be found from
$$I^2 (R_{ab} + R_{cd}) = (4)^2 (3)$$
$$= 48 \quad \text{watts}$$
In the motor the electrical energy is transformed into mechanical power, some which is lost in friction and magnetic losses.
Some of the electrical energy delivered to the motor is also
used to overcome the resistance of the motor windings and manifests as heat. Therefore, the power consumed by the resistance
of the motor is
$$I^2 R_{bc} = (16)(2)$$
$$= 32 \quad \text{watts}$$
and the mechanical power developed by the motor is
$$392.8 - 32 = 360.8 \quad \text{watts}$$
Or, since there are 746 watts in one horsepower, the mechanical
power developed by the motor is $\dfrac{360.8}{746} \cong \dfrac{1}{2}$ horsepower

We can check this computation finding the emf of the motor from
the potential across the motor less the voltage drop in the
resistance of the motor. Thus
$$\text{Emf (motor)} = 98.2 - 4(2)$$
$$= 90.2 \quad \text{volts}$$
Mechanical power delivered by the motor is, therefore,
$$\text{Emf (motor)} \times I \text{ (through the motor)}$$
hence the developed power is given by
$(90.2)(4) = 360.8$ watts or very nearly one-half horsepower.
Since some of this power is used to overcome friction and wind-
age and some other losses, the motor would probably deliver
about 0.45 horsepower which could be used to drive some other
mechanical device.

3.5 Equivalent Circuit Representation of the Sources of Electri-
cal Energy.

All electric circuits receive the electrical energy from one
or more non-electrical energy sources. The electrical energy
source converts mechanical, thermal, chemical or radiant energy into electrical form, it does not create the electrical energy. Alessandro Volta was responsible for the chemical to the
electrical type of conversion by his development of the voltaic

cell. A voltaic cell consists of two electrodes which are inserted in an electrolyte, generally a solution of acid, base or
salt in water. Whereas the metallic conductor transports its
charge by electrons, the charge in electrolytes is generally
transported by ions, which are present in the solution, both positive and negative. When a metal electrode is placed into an
electrolytic solution a difference of potential develops between
the electrode and the solution. Thus, in a voltaic cell a difference
of potential will develop between the two electrodes if they are
of different materials in a common electrolyte. However, the
electrodes may be dipped into two separate solutions which are
separated by a partition which allows the ions to permeat. The
chemical behavior of the voltaic cell is such that a potential difference or net emf develops between the two electrodes. The net
emf which develops is found to be a function of the materials
used in the voltaic cell. The emf is independent of whether or not
charge is being transported through the cell. A group of cells
may be added in series so that the polarities are additive resulting in a higher value for the emf. This formation is known
as a battery. Since the electrolyte, the electrodes and the leads
of a battery generally possess a certain amount of resistance
the terminal voltage is generally less than the total emf of the
voltaic cells when current is flowing through an external load.
The factors which result in an internal voltage drop tending to
decrease the terminal voltage are lumped together and referred
to as the internal resistance, R_i, of the battery. The equivalent
electrical circuit for a battery comprising an emf E, the battery
internal resistance R_i, is shown supplying electrical power to a
load in Figure 11. It should be noted that E, the emf, is independent of the current which flows but V_{12} or V_t, the terminal voltage depends on the current. Thus as I_t increases the voltage
drop across R_i increases and by Kirchhoff's voltage relation V_t
decreases, since

$$V_t = E - R_i I_t$$

In Figure 12, a graph has been plotted of, V_t, terminal voltage as
a function of I_t, the load current.

It is important to recognize that the equivalent circuit
attempts to specify the practical circuit by considering this
type of electrical energy source as a constant voltage source,
the emf, and a terminal voltage which changes with the load current; i.e., a physical voltage source can be looked upon as an
ideal source, a constant voltage independent of the current and
a resistor in series. The terminal voltage of a practical voltage

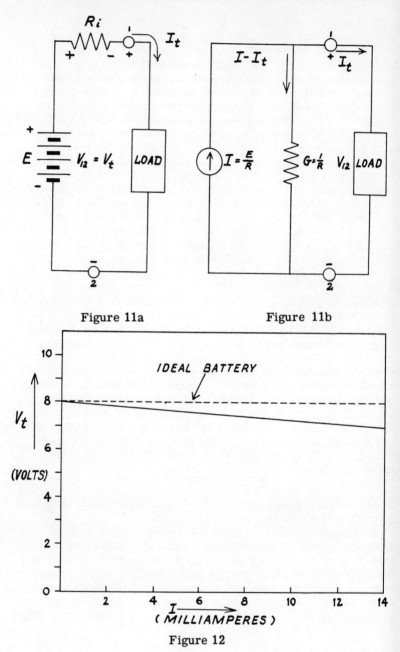

Figure 11a   Figure 11b

Figure 12

source is therefore always less than its emf by the amount of the voltage drop in the resistance.

All the other physical sources of electrical energy also employ actual physical apparatus, whether they are thermal, mechanical or radiant energy systems, and hence they always contain resistance. Therefore all physical sources produce a potential difference at the terminals less than the emf developed. They may all, therefore, be represented by the equivalent circuit described above, namely, a constant voltage, the emf, independent of the load current and a series resistor to account for the potential difference at the terminals of the source.

We shall see that in the next chapter, dealing with the simplification of electrical networks, it is sometimes more desirable to know the current of the source. It is important therefore to be able to convert a voltage source into an equivalent current source. Such an equivalent circuit is shown in Figure 11b. The current source, like the voltage source, must take into account the losses that accompany the conversion of the non electrical power into the electrical form. It does so by the use of an internal shunt or parallel conductance equal to the reciprocal of the series resistance of the voltage source. The equivalence is obtained from the application of Ohm's law. Beginning with the relationship of the voltage source:

$$V_t = E - R_i I_t$$

and solving for the current I_t we obtain:

$$I_t = \frac{E}{R_i} - \frac{V_t}{R_i}$$

which we may write as

$$I_t = I - G_i V_t$$

A critical examination of this equation shows us that the equivalent current source consists of a current I which is independent of the terminal voltage, V_t, and a current $G_i V_t$ which increases as the terminal voltage increases, thereby reducing the current, I_t, which is available to the load. This condition may be symbollically represented in a circuit diagram, as in Figure 11b, by a constant current source, $I = \frac{E}{R}$, possessing an internal shunt conductance, $G_i = \frac{1}{R_i}$, through which the current, $G_i V_t$ flows.

Of course, for identical conditions of voltage and current at the terminals 1-2, the terminal voltage derived for Fig. 11a can be equated to that derived from Figure 11b. Also, the terminal current which was obtained from Figure 11b can also be obtained from Figure 11a. Hence the current source can be equated to the voltage source or vice-versa, as long as we keep the conversion relations in mind:

$$R_i = \frac{1}{G_i}$$

$$I = \frac{E}{R_i}$$

$$E = \frac{I}{G_i}$$

At this time a word of caution is essential. Under certain conditions, the relations above equate current and voltage sources in so far as the terminal voltage and current are concerned as well as the amount of energy converted. This is not generally true. In our present derivation the amount of energy converted by the equivalent sources is not the same and our equivalence holds only for the terminal voltage and current.

3.6 Numerical Example 4

The circuit of Figure 12a represents two batteries supplying power to a load. The internal resistance of each battery is 0.1 ohm. The resistance of the connecting wires is negligible in this problem. It is desired to find the power dissipated by the load.

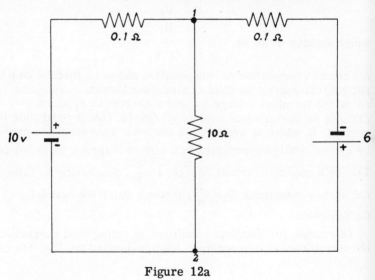

Figure 12a

Note that each battery is shown as a voltage source with a constant emf and a series resistance. We shall solve for the current in the load by converting the voltage sources to current
sources making use of the relations of section 3.5. The equivalent current sources are indicated in the equivalent circuit of
Figure 12b

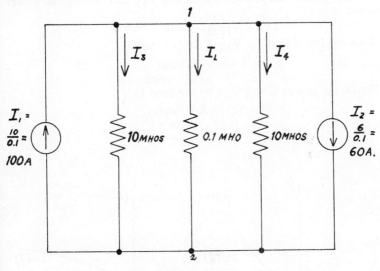

Figure 12b

The application of Kirchhoff's current law at node 1 yields the
following equation:
$$I_1 - I_3 - I_L - I_4 - I_2 = O$$
or $\qquad I_L = I_1 - I_2 - I_3 - I_4$

and from Ohm's law since $I = VG$ we have
$$I_L = 100 - 60 - V_{12}(10) - V_{12}(10)$$
but $\qquad I_L = V_{12}(0.1)$ by direct application of Ohm's law
to terminals 1-2.

Hence substituting $V_{12} = I_L(10)$ we obtain

$$I_L = 40 - I_L(100) - I_L(100)$$
or $\qquad (201) I_L = 40$

and $\qquad I_L = \dfrac{40}{201} \cong 0.20$ amperes

As a check, the student should solve the original circuit, with

the voltage sources, for the current in the load, I_L.
The power dissipated in the 10 ohm load is

$$P = I_L{}^2 R \quad \text{watts}$$
$$= (0.20)^2 \, 10$$
$$= 0.40 \quad \text{watts}$$

we may check this answer by finding V_{12} and finding the power
from

$$P = V_{12} I_L \quad \text{watts}$$

Here we found that $\quad V_{12} = 10 \, I_L \quad$ volts

$$= 10(0.2)$$
$$= 2 \quad \text{volts}$$

Hence the power dissipated by the load is

$$P = (2)(0.20) \quad \text{watts}$$
$$= 0.40 \quad \text{watts}$$

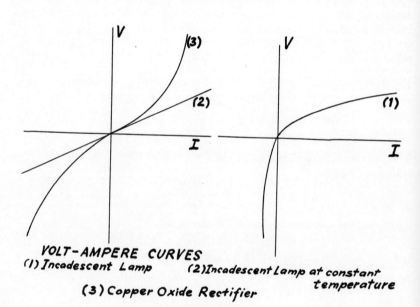

VOLT-AMPERE CURVES
(1) Incadescent Lamp   (2)Incadescent Lamp at constant temperature
(3) Copper Oxide Rectifier

Figure 4

CHAPTER 4—THE SIMPLIFICATION
OF ELECTRICAL CIRCUIT CALCULATIONS

The solution of electric circuit problems generally involves the application of the Kirchhoff's voltage and current laws and setting up a number of independent equations equal to the number of unknown quantities and solving these equation simultaneously. We have adopted the method of the simultaneous solution of these algebraic equations by the use of determinants. We have adopted the use of determinants in order to systematize our solution and to reduce the amount of computation which is involved. It is our desire to reduce the amount of labor involved in computation as much as possible. The adoption of a systematic procedure to set up the Kirchhoff equations so that the method of determinants for their solution is immediately applicable seems very much worthwhile.

Kirchhoff's basic laws may be re-stated in the following manner:
1. The sum of the voltage drops taken in a specified direction around any loop in a circuit equals the sum of the voltage rises in that specified direction.
2. The sum of the currents entering any node in a circuit is equal to the sum of the currents leaving that node. The laws when applied in these terms to electrical circuits yield two methods for the writing of the equations which tend to systematize the work involved and in complicated electrical networks some systematic method is essential in order to avoid confusion. The first law yields the so called loop or mesh method; while the second law yields the nodal method.

4.1 The Loop or Mesh Method of Circuit Analysis

We will consider the circuit of Figure 9, example 2, once more and this time apply the Kirchhoff's laws in the form stated above. We consider this circuit as made up of two loops or meshes a-b-d-c-a and c-d-f-e-c and we will write the voltage equations in the direction thus specified, namely, clockwise. From the circuit diagram we see that

$$I_{ab} = I_{bd}$$
$$I_{df} = I_{fe}$$

and that $\qquad I_{cd} = I_{fe} - I_{ab}$ at node d

Let $\qquad I_{ab} = I_1$ and $I_{fe} = I_2$

Then $\qquad I_{cd} = I_2 - I_1$

Thus we see that for a two loop network we need only specify two unknown currents. The voltage equations therefore become
For loop a-b-d-c-a

$$I_1 (0.1) + I_1 (1) - (I_2 - I_1) (0.1) = 115$$

and for the second loop c d f e c

$$(I_2 - I_1) (0.1) + I_2 (0.8) + I_2 (0.1) = 115$$

which reduce to the following two equations:

(1) $\qquad (0.1 + 1 + 0.1) I_1 - 0.1 I_2 \qquad\qquad = 115$

(2) $\qquad\qquad -(0.1) I_1 + (0.1 + 0.8 + 0.1) I_2 = 115$

We observe that in the first equation the coefficient of the I_1 term is composed of all the resistance in the first mesh, if that mesh were separated from the rest of the circuit. We make the same observation about the coefficient of I_2 in the second equation. We assign the name of the self resistance of a loop to such a term. We also observe that the coefficient of the I_2 term in equation (1) and the coefficient of the I_1 term in equation (2) are identical. This term represents the value of the resistance of the branch which couples loop a b d c a to loop c d f e c. This coupling is a direct one and the total resistance of the coupling branch is called the mutual resistance of the two loops. By studying the two equations carefully we can formulate a general rule for writing these loop equations directly. We shall do so in the next example. For the completion of our present problem we simplify the two equations and solve for the currents as follows:

$$(1) \quad 1.2 I_1 - 0.1 I_2 = 115$$
$$(2) \quad -0.1 I_1 + 1.0 I_2 = 115$$

hence

$$I_1 = \frac{\begin{vmatrix} 115 & -0.1 \\ 115 & 1.0 \end{vmatrix}}{\begin{vmatrix} 1.2 & -0.1 \\ -0.1 & 1.0 \end{vmatrix}} = \frac{126.5}{1.19} = 106.2 \text{ amperes}$$

$$I_2 = \frac{\begin{vmatrix} 1.2 & 115 \\ -0.1 & 115 \end{vmatrix}}{\begin{vmatrix} 1.2 & -0.1 \\ -0.1 & 1.0 \end{vmatrix}} = \frac{149.5}{1.19} = 125.5 \text{ amperes}$$

and $I_{cd} = I_2 - I_1 = 125.5 - 106.2$
$$= 19.3 \quad amperes$$
All of the answers check with the previous computation.
We will now consider a more complicated problem involving
three loops as shown in the circuit of Figure 13.

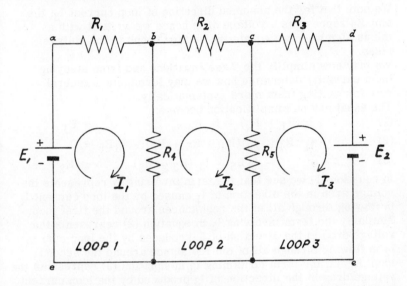

Figure 13

A careful study of the last example indicates that we may be able
to write the equations more systematically if we consider currents that circulate in each loop rather than the branch currents.
Although the assumed loop currents may be fictitious currents
we can readily determine the branch currents from the loop currents. The direction in which the loop currents circulate is arbitrarily assumed but by convention they are generally assumed
to flow in a clockwise direction. Thus the loop currents I_1, I_2
and I_3 are assumed to flow in the circuit of Figure 13. The
branch currents are readily found from these loop currents. Thus
$$I_{ab} = I_1 \; ; \; I_{bc} = I_2 \; ; \; I_{cd} = I_3$$
and
$$I_{be} = I_1 - I_2$$
$$I_{ce} = I_2 - I_3$$
Kirchhoff's voltage equation for the three loops in terms of the
loop currents may now be written:

(1) $I_1 R_1 + (I_1 - I_2) R_4 = E_1$

For loop b c e b we have:

(2) $I_2 R_2 + (I_2 - I_3) R_5 + (I_2 - I_1) R_4 = 0$

and for loop c d e c we have:

(3) $I_3 R_3 + (I_3 - I_2) R_5 = - E_2$

We note that for the assumed direction of loop current I_3, the emf E_2 represents a voltage drop hence we write it with a minus sign when we equate the voltage drops to the voltage rises.

We may now simplify the three equations and from studying them carefully determine how we may formulate a general rule for writing them more systematically.

The equations on simplification become:

(1) $I_1 (R_1 + R_4) - I_2 R_4 = E_1$

(2) $- I_1 R_4 + I_2 (R_4 + R_5 + R_2) - I_3 R_5 = 0$

(3) $- I_2 R_5 + I_3 (R_3 + R_5) = -E_2$

In equation (1) we see that the term involving I_1 represents the voltage drop in the direction of I_1 caused by the loop current I_1 in flowing through all of the resistances around the first loop. Similarly the term involving I_2 in equation (2) represents the voltage drop in the direction of I_2 produced by the loop current I_2 in flowing through all of the resistance around the second loop. Again, the term containing I_3 in equation (3) represents the voltage drop in the direction of I_3 produced by the loop current I_3 in flowing through all of the resistance around the third loop. By definition we now call all of the resistance around the contour of any loop, when it is severed from the rest of the circuit, as the self resistance of the loop. By definition, R_{11} is the total self resistance of the first loop; R_{22} is the total self resistance of the second loop and R_{33} represents the self resistance of loop 3.

The term involving I_2 in equation (1) can be seen to represent the voltage drop produced by I_2 flowing through all the resistances which are common to both loops one and two. We must keep in mind that we are applying Kirchhoff's voltage law to loop one in direction of the loop current, and writing the resulting equation as the sum of the voltage drops equal to the sum of the voltages rises in that direction. Therefore the voltage drop due to I_2 in flowing through the common resistances of loop one and two will be positive if it is in the direction of I_1. Here, of course, the voltage drop $I_2 R_4$ is positive in the direction opposite to that of I_1 and hence it will have a negative sign in front of it. The right hand side of all of the equations contain the net rise in emf in the respective loops in the direction of the loop currents.

We will carefully inspect once more the terms involving loop currents I_1 and I_3 in equation (2). Here we see that the term involving I_1 represents the voltage drop produced by I_1 flowing through all the resistances common to both loops one and two. Since this voltage drop is positive in the direction opposite to that of I_2 and since the equation (2) is written with respect to the voltage drops in the direction of I_2 the term $I_1 R_4$ is negative. Once again, the term involving I_3 in equation (2) represents the voltage drop produced by I_3 in flowing through all the resistance common to both loops two and three. Again since equation (2) represents voltage drop in the direction of I_2 and the voltage drop $I_3 R_5$ is positive in the direction opposite to I_2 it is written with a minus sign. All the resistances which are common to two branches are called mutual resistances. Thus R_4 is the resistance common to loops one and two and R_5 is the resistance common to loop two and three. The mutual resistance between loop one and two is referred to as R_{12} while the mutual resistance between loop two and three is R_{23}. It is obvious from the circuit that the mutual resistances between loops two and one $R_{21} = R_{12}$ while the mutual resistance between loops three and two, $R_{32} = R_{23}$. This relationship among the mutual resistances will be generally true for linear circuit elements which are common to two loops.

Using the definitions for self and mutual resistances we may now write the equations in the following general form:

(1) $\qquad R_{11} I_1 - R_{12} I_2 - R_{13} I_3 = E_1$

(2) $\qquad -R_{21} I_1 + R_{22} I_2 - R_{23} I_3 = 0$

(3) $\qquad -R_{31} I_1 - R_{32} I_2 + R_{33} I_3 = E_2$

where $\qquad R_{11} = (R_1 + R_4)$

$\qquad\qquad R_{22} = (R_2 + R_4 + R_5)$

$\qquad\qquad R_{33} = (R_4 + R_5)$

and $\qquad R_{12} = R_4 = R_{21}$

$\qquad\qquad R_{23} = R_5 = R_{32}$

$\qquad\qquad R_{13} = 0 \ \ = R_{31}$

For the systematic writing of the equation it is best to include all the terms even when they may be zero in the equations. The solution by determinants now readily follow:

$$I_1 = \frac{\begin{vmatrix} E_1 & -R_{12} & -R_{13} \\ O & R_{22} & R_{23} \\ -E_2 & -R_{32} & R_{33} \end{vmatrix}}{\begin{vmatrix} R_{11} & -R_{12} & -R_{13} \\ -R_{21} & R_{22} & -R_{23} \\ -R_{31} & -R_{32} & R_{33} \end{vmatrix}}$$

$$I_2 = \frac{\begin{vmatrix} R_{11} & E_1 & -R_{13} \\ -R_{21} & O & -R_{23} \\ -R_{31} & -E_2 & R_{33} \end{vmatrix}}{\begin{vmatrix} R_{11} & -R_{12} & -R_{13} \\ -R_{21} & R_{22} & -R_{23} \\ -R_{31} & -R_{32} & R_{33} \end{vmatrix}}$$

$$I_3 = \frac{\begin{vmatrix} R_{11} & -R_{12} & E_1 \\ -R_{21} & R_{22} & O \\ -R_{31} & -R_{23} & -E_2 \end{vmatrix}}{\begin{vmatrix} R_{11} & -R_{12} & -R_{13} \\ -R_{21} & R_{22} & -R_{23} \\ -R_{31} & -R_{32} & R_{33} \end{vmatrix}}$$

For the final solution we now substitute the values of R_{11}, R_{22}, R_{33}, the self resistances, and the values of R_{12}, R_{23}, R_{13}, the mutual resistances into the determinant equations. Since $R_{13} = R_{31} = 0$ the final equations for the currents in determinant form appears as:

$$I_1 = \frac{\begin{vmatrix} E_1 & -R_4 & O \\ O & (R_2 + R_4 + R_5) & -R_5 \\ -E_2 & -R_5 & (R_4 + R_5) \end{vmatrix}}{\begin{vmatrix} (R_1 + R_4) & -R_4 & O \\ -R_4 & (R_2 + R_4 + R_5) & -R_5 \\ O & -R_5 & (R_4 + R_5) \end{vmatrix}} \text{amperes}$$

$$I_2 = \frac{\begin{vmatrix} (R_1 + R_4) & E_1 & O \\ -R_4 & O & -R_5 \\ O & -E_2 & (R_4 + R_5) \end{vmatrix}}{\begin{vmatrix} (R_1 + R_4) & -R_4 & O \\ -R_4 & (R_2 + R_4 + R_5) & -R_5 \\ O & -R_5 & (R_4 + R_5) \end{vmatrix}} \text{ amperes}$$

$$I_3 = \frac{\begin{vmatrix} (R_1 + R_4) & -R_4 & E_1 \\ -R_4 & (R_2 + R_4 + R_5) & O \\ O & -R_5 & -E_2 \end{vmatrix}}{\begin{vmatrix} (R_1 + R_4) & -R_4 & O \\ -R_4 & (R_2 + R_4 + R_5) & -R_5 \\ O & -R_5 & (R_4 + R_5) \end{vmatrix}} \text{ amperes}$$

The final values for the currents are obtained from the expansion of the third order determinant as follows:

$$I_1 = \frac{E_1(R_2+R_4+R_5)(R_4+R_5)+E_1R_4R_5-E_1R_5^2}{(R_1+R_4)(R_2+R_4+R_5)(R_2+R_4)-R_5^2(R_1+R_4)-R_4^2(R_4+R_5)} \text{ amperes}$$

$$I_2 = \frac{-E_2(R_1+R_4)R_5+E_1(R_4+R_5)R_4}{(R_1+R_4)(R_2+R_4+R_5)(R_2+R_4)-R_5^2(R_1+R_4)-R_4^2(R_4+R_5)} \text{ amperes}$$

$$I_3 = \frac{-E_2(R_1+R_4)(R_2+R_4+R_5)+E_1R_4R_5+E_2R_4^2}{(R_1+R_4)(R_2+R_4+R_5)(R_2+R_4)-R_5^2(R_1+R_4)-R_4^2(R_4+R_5)} \text{ amperes}$$

A careful examination of the determinant form for the solution of the currents reveals the following characteristics:
1. As indicated in Cramer's rule for determinants all three denominators are identical and evaluate of course to the same constant.
2. The principal diagonal of the denominator determinant is made up of the following terms, R_{11}, R_{22} and R_{33}, i.e., all of the self resistances of the circuit.
3. A symmetry exists about this principal diagonal, since $R_{12} =$

R_{21}; $R_{23} = R_{32}$ and $R_{13} = R_{31}$. It is seen therefore that the symmetry exists about the principal diagonal and is made up of the mutual resistances, written with their proper algebraic signs. These characteristics may serve as a check in writing the equations and the solution need follow only after these characteristics are satisfactorily checked.

4.2 Numerical Example 5

We are required to find all the branch currents in the circuit of Figure 14.

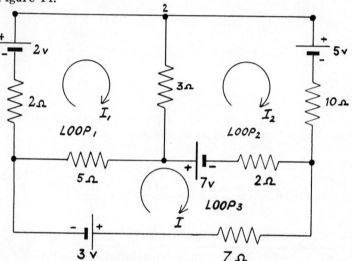

This circuit is seen to have 3 independent nodes and a total of 6 branches; hence by the rule specified in chapter 3 regarding the number of independent equations that are necessary to solve for the unknown currents is given by:

No. of independent equations = No. of branches - No. of ind. nodes
$$= 6 \qquad - \qquad 3$$
$$= 3$$

Hence we may choose the three obvious loops, specify the loop currents flowing in the clockwise direction in each loop and write the three independent loop equations

$$(1) \qquad R_{11}\,I_1\,-R_{12}\,I_2-\,R_{13}\,I_3 = E_1$$

$$(2) \qquad -R_{12}\,I_1 + R_{22}\,I_2 - R_{23}\,I_3 = E_2$$

$$(3) \qquad -R_{13}\,I_1 - R_{23}\,I_2 + R_{33}\,I_3 = E_3$$

where

$$R_{11} = (2 + 3 + 5) = 10 \text{ ohms}$$

$$R_{22} = (3 + 2 + 10) = 15 \text{ ohms}$$

$$R_{33} = (2 + 7 + 5) = 14 \text{ ohms}$$

$$R_{12} = R_{21} = 3 \text{ ohms}$$

$$R_{23} = R_{32} = 2 \text{ ohms}$$

$$R_{13} = R_{31} = 5 \text{ ohms}$$

$$E_1 = 2 \text{ volts}$$

$$E_2 = (7 - 5) = 2 \text{ volts}$$

$$E_3 = (-7 - 3) = -10 \text{ volts}$$

Hence the equations reduce to:

(1) $10 I_1 - 3 I_2 - 5 I_3 = 2$

(2) $-3 I_1 + 15 I_2 - 2 I_3 = 2$

(3) $-5 I_1 - 2 I_2 + 14 I_3 = -10$

and

$$I_1 = \frac{\begin{vmatrix} 2 & -3 & -5 \\ 2 & 15 & -2 \\ -10 & -2 & 14 \end{vmatrix}}{\begin{vmatrix} 10 & -3 & -5 \\ -3 & 15 & -2 \\ -5 & -2 & 14 \end{vmatrix}}$$

$$= \frac{380 - 674}{2040 - 499} = \frac{-294}{1541} = -0.190 \text{ amperes}$$

$$I_2 = \frac{\begin{vmatrix} 10 & 2 & -5 \\ -3 & 2 & -2 \\ -5 & -10 & 14 \end{vmatrix}}{1541}$$

$$= \frac{-16}{1541} \cong - 0.010 \text{ amperes}$$

and

$$I_3 = \frac{\begin{vmatrix} 10 & -3 & 2 \\ -3 & 15 & 2 \\ -5 & -2 & -10 \end{vmatrix}}{1541}$$

$$= \frac{-1178}{1541} = -0.760 \text{ amperes}$$

Therefore the branch currents are:

$$I_{12} = - 0.190 \quad \text{amperes}$$
$$I_{23} = - 0.180 \quad \text{amperes}$$
$$I_{24} = - 0.010 \quad \text{amperes}$$
$$I_{43} = + 0.750 \quad \text{amperes}$$
$$I_{31} = + 0.570 \quad \text{amperes}$$
$$I_{41} \text{ (through 7 ohm)} = - 0.760 \quad \text{amperes}$$

As a check apply Kirchhoff's current law at node 3, thus:

$$I_{31} + I_{32} + I_{34} = 0$$

or

$$0.570 + 0.180 - 0.750 = 0$$
$$0.750 - 0.75 = 0$$

4.3 The Node-Voltage Method of Electrical Circuit Analysis

In the loop or mesh method of circuit analysis only voltage equations, involving loop or mesh currents, were written. The voltage equations were solved simultaneously for the loop currents and the individual branch currents were then easily obtained. The node-voltage method demonstrates the technique of analyzing an electrical circuit in terms of potential differences between nodes or junction points whose values will be found from the solution of a set of simultaneous linear equations which are gotten from application of Kirchhoff's current law. Since the number of independent node equations is one less than the number of nodes, one node is arbitrarily chosen as a reference potential and is generally the ground or chassis connection. The reference node is generally assigned a value of zero potential, the other nodes are assigned numbers, 1, 2, 3, etc. and the potential difference between any two nodes is designated as V_1, $-V_2$. Therefore the potential difference between any node and the reference node will be designated as $V_1 - 0$ or simply V_1, $V_2 - 0$ or V_2 etc.

After designating the various nodes as above, some of the potential differences will of course be known, the rest will be found from the solution of the simultaneous equations that will be written from the application of Kirchhoff's current law. We have already indicated that the form in which we will apply the current law will be to equate the sum of the currents entering a node to the sum of the currents leaving the node. Also, since Ohm's law in the form

$$I = GV$$

most easily expresses a current in terms of a potential difference we will designate the dissipative circuit element as conductance rather than resistance. We will also simplify our analysis if we convert all voltage sources to current sources in accordance with section 3.5 of chapter 3.

Consider the circuit of Figure 15a.

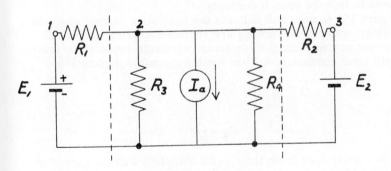

Figure 15a

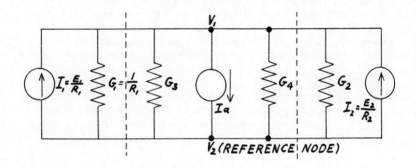

Figure 15b

It is assumed that the voltage sources are known; that the current source I_a and all the resistances are known. It is desired to find the branch currents.

Figure 15b is drawn to indicate the equivalent circuit when the voltage sources E_1 and E_2 are replaced by their equivalent current sources and the resistance elements replaced by their equivalent conductance. Thus for the circuit of Figure 15b

$$I_1 = \frac{E_1}{R_1} \quad ; \quad G_1 = \frac{1}{R_1} \quad ; \quad G_3 = \frac{1}{R_3}$$

$$I_2 = \frac{E_2}{R_2} \quad ; \quad G_2 = \frac{1}{R_2} \quad ; \quad G_4 = \frac{1}{R_4}$$

It is readily seen from the circuit diagrams that the circuit of Figure 15a has 4 nodes or 3 independent nodes and hence requires three independent voltage equations to solve for all the unknown quantities. By conversion to the equivalent current sources as indicated in Figure 15b we have reduced the circuit to only two nodes so that there is only one independent node and hence only one independent equation to solve. This is indeed a great simplification. We will make node 2 the reference node and apply the Kirchhoff current law to node 1. We obtain:

$$(V_1 - 0) G_1 + (V_1 - 0) G_3 + (V_1 - 0) G_4 + (V_1 - 0) G_2 + I_a = I_1 + I_2$$

which on simplification becomes:
$$V_1 (G_1 + G_2 + G_3 + G_4) = I_1 + I_2 - I_a$$

Thus Current entering   =   Current leaving
  Hence
  V_1 is now easily evaluated

$$V_1 = \frac{I_1 + I_2 - I_a}{G_1 + G_2 + G_3 + G_4}$$

and the branch currents readily follow. We shall now set up the systematic method of nodal analysis for the solution of electrical circuits. We now apply this method to the circuit of Figure 16, where the currents specified are known as are all the conductances.

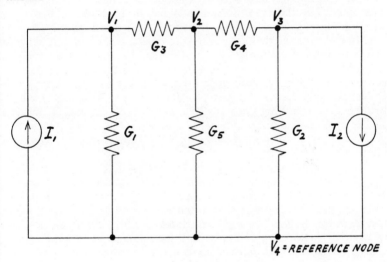

Figure 16

It is desired to find the branch currents. By inspection of the circuit diagram we note that there are 4 nodes or 3 independent nodes. Arbitrarily we choose node 4 as the reference node and assign the potentials of the 3 nodes, relative to the zero or reference node, as V_1, V_2 and V_3. We now apply the Kirchhoff current law to each of the 3 nodes.

Node 1: $V_1 G_1 + (V_1 - V_2) G_3$ $\qquad\qquad\qquad\qquad = I_1$

Node 2: $(V_2 - V_1) G_3 + V_2 G_5 + (V_2 - V_3) G_4$ $\qquad\quad = 0$

Node 3: $(V_3 - V_2) G_4 + V_3 G_2$ $\qquad\qquad\qquad\qquad = -I_2$

These equations may be written in the following manner:

    Currents Entering $\qquad = \qquad$ Currents Leaving

Node 1: $V_1 (G_1 + G_3) - V_2 G_3$ $\qquad\qquad\qquad\qquad = I_1$

Node 2: $-V_1 G_3 \qquad + V_2 (G_3 + G_5 + G_4) -V_3 G_4$ $\qquad = 0$

Node 3: $\qquad\qquad\quad - V_2 G_4 \qquad\qquad + V_3 (G_4 + G_2) \quad = -I_2$

The equations have been arranged in this fashion to facilitate the solution by means of determinants and to bring out clearly certain relations which will yield a systematic manner of writing the equations. Thus, the term associated with V_1 in equation 1 is seen to be the sum of the conductances of the branches which terminate upon node 1; the term associated with V_2 in equation 2 represents the sum of the conductances of the branches which terminate upon node 2; and finally, the coefficient of V_3 in equation 3 is seen to be the sum of the conductances of the branches which terminate upon node 3. Hence we may define the self-conductance of a node, as the total conductance of the branches terminating upon it. Thus $G_{11} = G_1 + G_3$ and is defined as the self conductance of node 1; $G_{22} = G_3 + G_5 + G_4$ is the self conductance of node 2 and $G_{33} = G_4 + G_2$ is the self conductance of node 3.

Similarly, it may be seen that the coefficient of V_2 in equation 1 represents the mutual conductances between node 1 and node 2 when all the remaining nodes (here, of course, only node 3) have been shorted directly to the reference node. We note that the term has a negative sign associated with it, since we have written an equation which specifies the current as flowing from node 2 to node 1. The mutual conductances between any two nodes, such as node 1 and the remaining nodes, except the reference node, are defined as the conductances of the branches joining those two nodes when all the remaining nodes have been shorted directly to the reference node. Thus, $G_{12} = G_3$ and is the mutual conductance between node 1 and node 2; $G_{23} = G_4$ and is the mutual conductance between node 2 and node 3 and $G_{13} = 0$ (in this circuit) represents the mutual conducatance between nodes 1 and 3. We may therefore write these equations in a very general way:

$$\text{Node 1: } V_1 G_{11} - V_2 G_{12} - V_3 G_{13} = I_1$$

$$\text{Node 2: } -V_1 G_{21} + V_2 G_{22} - V_3 G_{23} = 0$$

$$\text{Node 3: } -V_1 G_{31} - V_3 G_{32} + V_3 G_{33} = -I_2$$

It is important to observe the similarity of the nodal equations to the loop equations. The mutual terms are all written with a negative sign. Also the symmetry that we found for the loop equations exists here too since $G_{12} = G_{21}$ and represent the same circuit element. The solution for V_1, V_2 and V_3 readily follow from the application of the Cramer Rule:

$$V_1 = \frac{\begin{vmatrix} I_1 & -G_{12} & -G_{13} \\ 0 & G_{22} & -G_{23} \\ -I_2 & -G_{32} & G_{33} \end{vmatrix}}{\begin{vmatrix} G_{11} & -G_{12} & -G_{13} \\ -G_{21} & G_{22} & -G_{23} \\ -G_{31} & -G_{32} & G_{33} \end{vmatrix}} \text{ (volts)}$$

$$V_2 = \frac{\begin{vmatrix} G_{11} & I_1 & -G_{13} \\ -G_{21} & 0 & -G_{23} \\ -G_{31} & -I_2 & G_{33} \end{vmatrix}}{\begin{vmatrix} G_{11} & -G_{12} & -G_{13} \\ -G_{21} & G_{22} & -G_{23} \\ -G_{31} & -G_{32} & G_{33} \end{vmatrix}} \text{ (volts)}$$

$$V_3 = \frac{\begin{vmatrix} G_{11} & -G_{12} & I_1 \\ -G_{21} & G_{22} & 0 \\ -G_{31} & -G_{23} & -I_2 \end{vmatrix}}{\begin{vmatrix} G_{11} & -G_{12} & -G_{13} \\ -G_{21} & G_{22} & -G_{23} \\ -G_{31} & -G_{32} & G_{33} \end{vmatrix}} \text{ (volts)}$$

Here, as in the solution for the loop currents from the loop equations, we observe that the principal diagonal of the determinant of the denominator is composed of the <u>self</u> conductances, all with positive signs. The <u>mutual</u> conductances are seen to be symmetrically distributed about the main diagonal from the upper left end corner to the lower right hand corner. That is, the elements which occupy corresponding positions equidistant from the principal diagonal and on opposite sides of it are identi-

cal. It is also important to observe that the mutual conductances
all have negative signs.

4.4 Numerical example 6

We shall solve the circuit shown in Figure 17a by the application of the nodal-voltage method for the power dissipated in the
10 ohm resistor.

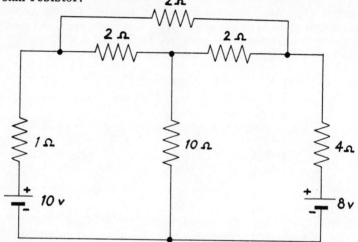

First, we simplify the circuit by replacing the voltage sources
by their equivalent current sources and the resistances by their
equivalent conductances with the resulting diagram shown in
Figure 17b

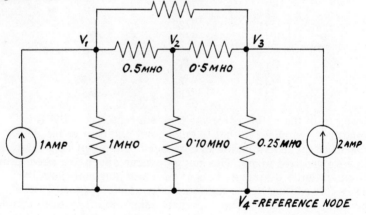

After circuit simplification we see that there are 4 nodes and therefore 3 independent current equations are required. The nodes are assigned to potentials V_1, V_2 and V_3 with respect to the reference potential V_4 (which is arbitrarily made equal to zero.) The current equations are now written with respect to the three nodes.

Node 1: $V_1 G_{11} - V_2 G_{12} - V_3 G_{13} = 1$

Node 2: $-V_1 G_{12} + V_2 G_{22} - V_3 G_{23} = 0$

Node 3: $-V_1 G_{13} - V_2 G_{23} + V_3 G_{33} = 2$

where

$$G_{11} = (1 + 0.5 + 0.5) \quad\; = 2$$

$$G_{22} = (0.5 + 0.5 + 0.10) = 1.1$$

$$G_{33} = (0.5 + 0.5 + 0.25) = 1.25$$

$$G_{12} = G_{21} = 0.5$$

$$G_{23} = G_{32} = 0.5$$

$$G_{13} = G_{31} = 0.5$$

The equations, therefore, evaluate to

$$2 V_1 - 0.5 V_2 - 0.5 V_3 \;\; = 1$$

$$- 0.5 V_1 + 1.1 V_2 - 0.5 V_3 \;\; = 0$$

$$- 0.5 V_1 - 0.5 V_2 + 1.25 V_3 = 2$$

and the solution is given below:

$$V_1 = \frac{\begin{vmatrix} 1 & -0.5 & -0.5 \\ 0 & 1.1 & -0.5 \\ 2 & -0.5 & 1.25 \end{vmatrix}}{\begin{vmatrix} 2 & -0.5 & -0.5 \\ -0.5 & 1.1 & -0.5 \\ -0.5 & -0.5 & 1.25 \end{vmatrix}} = \frac{2.725}{1.4125} = 1.93 \text{ volts}$$

$$V_2 = \frac{\begin{vmatrix} 2 & 1 & -0.5 \\ -0.5 & 0 & -0.5 \\ -0.5 & 2 & 1.25 \end{vmatrix}}{1.4125} = \frac{3.375}{1.4125} = 2.38 \text{ volts}$$

$$V_3 = \frac{\begin{vmatrix} 2 & -0.5 & 1 \\ -0.5 & 1.1 & 0 \\ -0.5 & -0.5 & 2 \end{vmatrix}}{1.4125} = \frac{4.70}{1.4125} = 3.34 \text{ volts}$$

As a check on the computation evaluate the current law about node 1:

$$2(1.93) - 0.5(2.38) - 0.5(3.34) = 1$$
$$3.86 - 1.19 - 1.67 \qquad\qquad = 1$$
$$3.86 - 2.86 \qquad\qquad = 1 \quad \text{check.}$$

The current flowing through the 10 ohm resistor is easily found. It is given by:

$$I_{24} = V_2 \, G$$
$$= 2.38 \, (0.10)$$
$$= 0.238 \text{ amperes}$$

The power dissipated in the 10 ohm resistor is given by:

$$P = I^2 \, R \qquad\qquad \text{watts}$$
$$= (0.238)^2 \text{ x } 10 \quad \text{watts}$$
$$= 0.575 \qquad\qquad \text{watts}$$

CHAPTER 5—ELECTRICAL NETWORK THEOREMS

Many electrical circuit theorems have been formulated that
are particularly useful for the solution of certain electrical networks. In many complicated circuits, we would like to know
whether any short-cuts are available to us for the computation
of a partial solution which may be desirable. Again, at times in
our study of electric circuits, the details of a certain section of
a circuit may be of no particular interest, and as long as we
can properly take into account the effect of such a section on the
whole circuit we will be quite satisfied. Sometimes we prefer to
think of the section of the circuit as simply a number of terminals, say 2, 3 or 4 and we would like to be in a position to replace
the section of the circuit by the simplest possible circuit joining the terminals. Hence a pair of terminals can be considered
an equivalent circuit provided that the same voltage relations
exists between the terminals as do in the original circuit for the
given currents flowing in and out of these terminals. Also, of
course, the total electrical power flowing into and out of the
equivalent circuit must be the same as that flowing into and out
of the actual circuit. The distribution of the power within the
two networks need not, however, be the same. We have already
discussed such equivalent circuits in Chapter 3 where we learned how to convert voltage sources into equivalent current
sources. Here we also learned to reduce series and parallel
combinations of resistances and conductances to the simplest
form of a single resistance and single conductance element. We
wish now to consider some additional equivalent circuits which
will be extremely useful in simplifying the electrical circuit calculations.

5.1 Equivalent Three Terminal Networks Containing No Electro-motive Forces.

Resistances in various combinations often occur in circuits
between three terminals instead of two terminals. Any three
terminal network, containing no emfs, can be reduced to one of
the two networks shown in Figures 18a, 18b, 18c and 18d

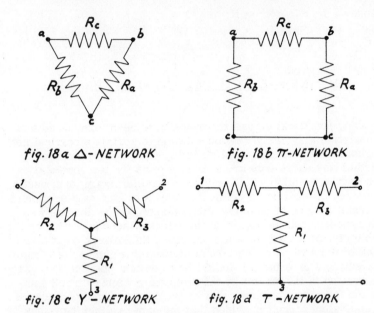

fig. 18a △-NETWORK

fig. 18b π-NETWORK

fig. 18c Y-NETWORK

fig. 18d T-NETWORK

Common arrangements of these circuits are called delta (Δ) or pi (π) circuits and wye (Y) or tee (T) circuits. The wye and delta are frequently found in electrical power publications while the pi (π) and tee (T) is generally the common terminology of the communication technologist. The T and Y are obviously the same circuit; also the π and Δ represent the same circuit. Many times difficult circuit arrangements can be greatly simplified by the conversion of the wye circuit to a delta circuit. The relations for conversion may be obtained as follows:

The resistance between terminals a and c in Figure 18a must be identical with the resistance between terminals 1 and 3 in Figure 18c or

$$\frac{R_b \ (R_a + R_c)}{R_a + R_b + R_c} = R_1 + R_2 \qquad \text{Eq. 5.1}$$

Similarly

$$\frac{R_c \ (R_a + R_b)}{R_a + R_b + R_c} = R_2 + R_3 \qquad \text{Eq. 5.2}$$

and

$$\frac{R_a \ (R_b + R_c)}{R_a + R_b + R_c} = R_1 + R_3 \qquad \text{Eq. 5.3}$$

Hence, when R_a, R_b and R_c are known R_1, R_2 and R_3 can be found, since there are 3 equations relating 3 unknowns. Similarly, if R_1, R_2 and R_3 are known R_a, R_b and R_c can be found. Say that R_a, R_b and R_c are known, that is, we are given the delta or π and we desire to find the equivalent wye or T. We may rewrite the 3 equations as follows:

$$1 R_1 + 1 R_2 + 0 R_3 = \frac{R_b (R_a + R_c)}{R_a + R_b + R_c} = a$$

$$0 R_1 + 1 R_2 + 1 R_3 = \frac{R_c (R_a + R_b)}{R_a + R_b + R_c} = b$$

$$1 R_1 + 0 R_2 + 1 R_3 = \frac{R_a (R_b + R_c)}{R_a + R_b + R_c} = c$$

We may now solve for R_1, R_2 and R_3 by the use of determinants.

$$R_1 = \frac{\begin{vmatrix} a & 1 & 0 \\ b & 1 & 1 \\ c & 0 & 1 \end{vmatrix}}{\begin{vmatrix} 1 & 1 & 0 \\ 0 & 1 & 1 \\ 1 & 0 & 1 \end{vmatrix}} = \frac{a+c-b}{2} = \frac{R_b R_c}{R_a+R_b+R_c}$$

where $a+b-c = \dfrac{R_bR_a+R_bR_c-R_aR_c-R_cR_b+R_a+R_b+R_aR_c}{R_a+R_b+R_c}$

$$= \frac{2 R_a R_b}{R_a + R_b + R_c}$$

$$R_2 = \frac{\begin{vmatrix} 1 & a & 0 \\ 0 & b & 1 \\ 1 & c & 1 \end{vmatrix}}{2} = \frac{b+a-c}{2} = \frac{R_b R_c}{R_a+R_b+R_c}$$

and
$$R_3 = \frac{\begin{vmatrix} 1 & 1 & a \\ 0 & 1 & b \\ 1 & 0 & c \end{vmatrix}}{2} = \frac{c+b-a}{2}$$

$$= \frac{R_a R_b}{R_a + R_b + R_c}$$

Therefore when the delta or π is known the wye or T can be found from the following equations:

Eq. 5.4
$$R_1 = \frac{R_a R_b}{R_a + R_b + R_c}$$

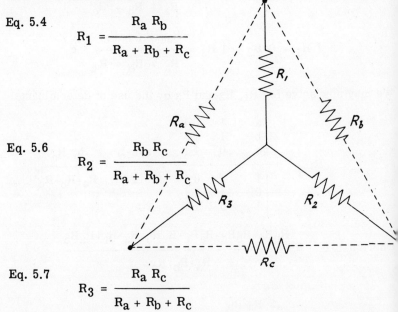

Eq. 5.6
$$R_2 = \frac{R_b R_c}{R_a + R_b + R_c}$$

Eq. 5.7
$$R_3 = \frac{R_a R_c}{R_a + R_b + R_c}$$

Figure 19

To solve for R_a, R_b and R_c when R_1, R_2 and R_3 are given, that is to find the equivalent π or delta for a known wye or T we make use of equations 5.4, 5.6 and 5.7 as follows:

Form the sum
$$R_1 R_2 + R_2 R_3 + R_3 R_1 = \frac{R_a R_b R_c}{R_a + R_b + R_c}$$

and divide this sum by each of the equations 5.4, 5.5 and 5.6 yielding the desired relations. Thus:

$$R_c = \frac{R_a\,R_b\,R_c}{R_a + R_b + R_c} \cdot \frac{R_a + R_b + R_c}{R_a\,R_b} = \frac{R_1\,R_2 + R_2\,R_3 + R_3\,R_1}{R_1}$$

$$R_a = \frac{R_a\,R_b\,R_c}{R_a + R_b + R_c} \cdot \frac{R_a + R_b + R_c}{R_a\,R_c} = \frac{R_1\,R_2 + R_2\,R_3 + R_3\,R_1}{R_2}$$

$$R_b = \frac{R_c\,R_b\,R_c}{R_a + R_b + R_c} \cdot \frac{R_a + R_b + R_c}{R_a\,R_c} = \frac{R_1\,R_2 + R_2\,R_3 + R_3\,R_1}{R_3}$$

and the following equations may be used to convert a wye or T to an equivalent delta or π:

$$R_a = \frac{R_1\,R_2 + R_2\,R_3 + R_3\,R_1}{R_2} \qquad \text{Eq. 5.6}$$

$$R_b = \frac{R_1\,R_2 + R_2\,R_3 + R_3\,R_1}{R_3} \qquad \text{Eq. 5.7}$$

$$R_c = \frac{R_1\,R_2 + R_2\,R_3 + R_3\,R_1}{R_1} \qquad \text{Eq. 5.8}$$

If the geometrical arrangement is drawn as is shown in Figure 19 a scheme for conversion is possible. Thus, each resistance in the equivalent wye is obtained by taking the product of the two delta resistances on each side of it and dividing by the sum of the delta resistances. Similarly, each resistance in the equivalent delta is obtained by taking the sum of the products of the wye resistances two at a time and dividing by the wye resistance opposite the desired delta resistance.
Consider the network shown in Figure 20a

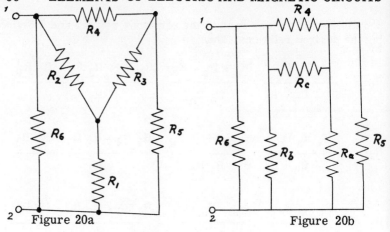

Figure 20a

Figure 20b

We wish to find the resistance between terminals 1 and 2 when
$R_1 = R_2 = R_3 = R_4 = R_5 = R_6 = 2$ ohms
Figure 20b shows the resulting simplified circuit, when the wye
is converted to a delta

$$R_a = \frac{R_1 R_2 + R_2 R_3 + R_3 R_1}{R_2} = \frac{4 + 4 + 4}{2} = 6 \text{ ohms}$$

and since $R_1 = R_2 = R_3 = 2$ ohms

$$R_a = R_b = R_c = 6 \text{ ohms}$$

The resistance between terminals 1 and 2 is now easily calcu-
lated since the parallel combination of 2 ohms and 6 ohms equals
1.5 ohms.
Hence:

$$R_{12} = \frac{1.5 \,(1.5 + 1.5)}{1.5 + 1.5 + 1.5} = \frac{4.5}{4.5} = 1 \text{ ohm}$$

5.2 The Superposition Theorem

In order to appreciate the principal of superposition as it
applies to linear circuits we will consider an example of a circuit with several sources present. The principle of superposition tells us that we may find the circuit response to each source,

acting alone in the circuit and finally adding the various responses
to find the total response of the circuit to all the sources. Figure
21 represents a two mesh network with a source in each loop. It
is desired to find the branch currents. The resistance of each
source indicated by a total series resistor. Thus the source resistances are each 1.5 ohms.

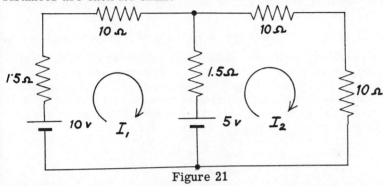

Figure 21

We shall first solve this simple circuit by assuming the loop
currents I_1 and I_2 as shown in the figure. The loop current
equations are:

loop 1: $I_1 (10 + 1.5 + 1.5) - (1.5) I_2 = 10 - 5$

loop 2: $-I_1 (1.5)   + I_2 (10 + 10 + 1.5) = 5$

On simplification we obtain:

$$13 I_1 - 1.5 I_2 = 10 - 5$$

$$-1.5 I_1 + 21.5 I_2 = 5$$

and solving for I_1 we get:

$$I_1 = \frac{\begin{vmatrix} (10-5) & -1.5 \\ 5 & 21.5 \end{vmatrix}}{\begin{vmatrix} 13 & -1.5 \\ -1.5 & 21.5 \end{vmatrix}} = \frac{(10-5)(21.5)}{276.25} + \frac{5(1.5)}{276.25}$$

$$= \frac{(10) \, 21.5}{216.25} + \frac{(5)(21.5)}{276.25} + \frac{(5)(1.5)}{276.25}$$

$$= \quad 10 \left(\frac{21.5}{276.25}\right) - 5 \left(\frac{20}{276.25}\right) \quad \text{amperes}$$

The answer to I_1 has been deliberatly obtained in this form to illustrate that both the 10 volt source and the 5 volt source contribute to the total current. Thus, if the 5 volt battery were omitted from the circuit the current I_1' that would flow would be

$$I_1' = \quad 10 \left(\frac{21.5}{276.25}\right) \quad \text{amperes and, it must be kept in mind,}$$

that although the source voltage was removed by letting it equal zero, the resulting current, I_1', is computed on the basis that the internal resistance of the 5 volt source is still in the circuit.

Similarly, if the 10 volt source were removed or set equal to zero, while its internal resistance is left in the circuit, the resulting current I_1'' which flows is:

$$I_1'' = \quad - 5 \left(\frac{20}{276.25}\right) \quad \text{amperes}$$

Thus, if we had removed the first voltage source from the circuit and replaced it with a resistance equal to its internal resistance, the current I_1'' would have resulted in mesh one. If we had removed the second source from the circuit and replaced it with a resistance equal to its internal resistance, the resultant loop current in mesh one would have been I_1'. Of course the total loop current $I_1 = I_1' + I_1''$ and flows in mesh one when both sources are present. We may conclude that each emf produces its own current and the actual resultant current is the algebraic sum of the individual currents. The superposition principle states therefore, that in any circuit consisting of generators and linear resistances, the current flowing at any point in the circuit is the sum of the currents that would flow if each generator were considered separately, all other generators being replaced at the tim by resistances equal to their internal resistances.

We will now complete the problem by applying the superposition principle. First, we shall consider the circuit of Figure 21 when the 5 volt source is short-circuited so that its emf in the circuit is equal to zero and it is simply replaced by a resistance of 1.5 ohms, which is the internal resistance of this source when it is active in the circuit. We may now write the voltage equations for

loop 1: $13 \ I_1' - 1.5 \ I_2' = 10$

loop 2: $-1.5 \ I_1' + 21.5 \ I_2' = 0$

where I_1' and I_2' are the loop currents which flow when the 5 volt source is removed from the circuit. The simultaneous solution of these equations yields:

$$I_1' = \frac{\begin{vmatrix} 10 & -1.5 \\ 0 & 21.5 \end{vmatrix}}{\begin{vmatrix} 13 & -1.5 \\ -1.5 & 21.5 \end{vmatrix}} = \frac{215}{276.25} = 0.78 \text{ amperes}$$

$$I_2' = \frac{\begin{vmatrix} 13 & 10 \\ -1.5 & 0 \end{vmatrix}}{276.25} = \frac{15}{276.25} = 0.054 \text{ amperes}$$

Now, considering only the 5 volt source present in the circuit, after short-circuiting the volt source and replacing it by a resistance of 1.5 ohms equal to its internal resistance, the voltage equations become:

loop 1: $13\, I_1'' - 1.5\, I_2'' = -5$

loop 2: $-1.5\, I_1'' + 21.5\, I_2'' = 5$

where I_1'' and I_2'' are the loop currents which flow when the 10 volt source is removed from the circuit.
The simultaneous solution of these equations yields:

$$I_1'' = \frac{\begin{vmatrix} -5 & -1.5 \\ 5 & 21.5 \end{vmatrix}}{\begin{vmatrix} 13 & -1.5 \\ -1.5 & 21.5 \end{vmatrix}} = \frac{-100}{276.25} = -0.363 \text{ amperes}$$

$$I_2'' = \frac{\begin{vmatrix} 13 & -5 \\ -1.5 & 5 \end{vmatrix}}{276.25} = \frac{57.5}{276.25} = 0.208 \text{ amperes}$$

The total loop currents flowing in the circuit when both sources are present is found from the algebraic summation of the cur-

rents which flow when each source is considered separately.
Thus the total loop current for mesh 1 is

$$I_1 = I_1' + I_1''$$

$$= 0.780 - 0.363$$

$$= 0.417 \quad \text{amperes}$$

The total loop current for mesh 2 is

$$I_2 = I_2' + I_2''$$

$$= 0.054 + 0.208$$

$$= 0.262 \quad \text{amperes}$$

The current through the mutual branch is, of course, $I_1 - I_2 =$
0.155 amperes.
We may now check these results by computing the total loop currents directly.

Thus, we have already found that the loop current, I_1, when
both sources are active together in the circuit is given by

$$I_1 = \frac{10 \ (21.5)}{276.25} - 5 \left(\frac{20}{276.25} \right)$$

$$= 0.780 - 0.363$$

$$= 0.417 \text{ amperes} \quad \text{(check)}$$

The loop current I_2 is given by:

$$I_2 = \frac{\begin{vmatrix} 13 & 5 \\ -1.5 & 5 \end{vmatrix}}{\begin{vmatrix} 13 & -1.5 \\ -1.5 & 21.5 \end{vmatrix}} = \frac{72.5}{276.25} = 0.263 \text{ amperes (check)}$$

The superposition principle is based upon the mathematical
form of the original equations. We have, of course, assumed
that all the resistances were independent of the current flowing
through them. That is, the volt-ampere curve for each resistor
is a straight line curve passing through the origin. If the curve
is not a straight line, the resistor will be non linear and the

superposition theorem does not apply. The principle of superposition represents one of the short-cuts of circuit analysis. It
becomes especially useful when we wish to find the effect of
only one particular emf in a circuit containing several sources.
It is important to note that the several sources considered and
illustrated were voltage sources. Of course the superposition
principle for linear circuits applies equally as well to current
sources. In the application of the principle it was demonstrated
how a voltage source may be properly removed. A current
source may be properly removed by open-circuiting the two
points in the circuit to which it is connected. Since a current
source has its internal conductance in shunt, open-circuiting
the current source leaves the internal shunt conductance in the
circuit as is required in the proper application of this theorem.

5.3 Thevenin's Theorem

One of the most useful short-cuts in electrical circuit analysis is attributed to M. L. Thevenin, a French scientist who re-
ported his findings in 1883. It enables us to find an equivalent
circuit for a two terminal network containing emf's. The
theorem is quite general and applies to alternating as well as
direct current circuit analysis. For our use, it may be stated
in the following form. Any network of resistance elements and
voltage sources if viewed from any two points in the network
may be replaced by a voltage source and a resistance in series
between these two points. Figures 22a and 22b illustrate the
statement. Consider any network, containing emf's, confined to
the box illustrated in Figure 22a from which the two terminals
a and b have been brought out. Figure 22b represents the
Thevenin equivalent network.

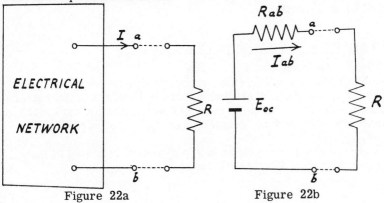

Figure 22a   Figure 22b

In order to use the theorem the equivalent Thevenin source
E_{OC} and the equivalent series resistance R_{ab} must be found.
These two terms represent the open circuit voltage drop V_{ab}
at the two terminals a and b and R_{ab}, the resistance at the two
terminals a and b when no current flows out of or into the terminals, and all the emfs have been properly removed from the
circuit. Thus, after bringing out the two terminals, a and b, the
original voltage drop across the terminals is computed or
measured. Then, all the emf's are properly removed from the
circuit, i.e., voltage sources are short-circuited and replaced
by resistances equal to their series internal resistance; current
sources are open-circuited and their internal shunt conductances are retained in the circuit by equivalent shunt conductances placed across the open circuited terminals. The resistance
across the open terminals a and b, R_{ab}, is then found either by
measurement or computation. This resistance represents the
net resistance of the original network between terminals a and
b.

The equivalent Thevenin circuit really represents a voltage
source, which may then be connected to any resistance load to
determine the current through the load. Thus if a resistance R
is now connected across terminals a and b the current flowing
in R will be given by

$$I_{ab} = \frac{E_{OC}}{R_{ab} + R} = amperes$$

We will now demonstrate the method of obtaining the equivalent
Thevenin voltage source in accordance with the above state-
ments. We will also be able to demonstrate that this equivalent
circuit yields the correct circuit response and hence may be
used as an equivalent circuit.

5.4 Applications of Thevenins' Theorem

Consider the circuit of Figure 23a
We shall obtain the Thevenin equivalent voltage source at the
terminals a-b. Since there is no current leaving or entering terminals a-b there will be no current flowing through the branch ac.
Hence the voltage drop from a to b is $V_{ac} + V_{cb}$ and since $V_{ac} =$
0, $V_{ab} = V_{cb}$.
Assume a loop current flows in the direction indicated in the
figure; then from Kirchhoff's voltage law

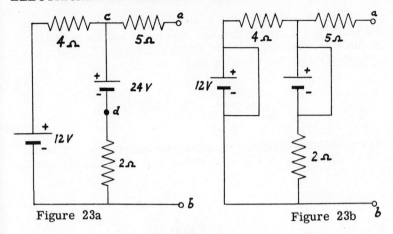

Figure 23a   Figure 23b

$$12 - 6I - 24 = 0$$

and
$$I = \frac{-12}{6} = -2 \text{ amperes}$$

The minus sign means that the actual current flows from node b to node c producing a voltage drop from b to d:

$$V_{bd} = I_{bd} R$$

$$= 2(2) = 4 \text{ volts}$$

Hence
$$V_{ab} = 24 - 4$$

$$= 20 \text{ volts}$$

But the Thevenin open-circuit voltage, E_{oc}, at the terminals a-b is V_{ab}; hence

$$E_{oc} = V_{ab} = 20 \text{ volts}$$

The equivalent series resistor for the Thevenin voltage source is obtained by computing the resistance across the terminals from a to b when the two voltage sources are short circuited. We may assume in this problem that the 4 ohms resistor in series with the 12 volt source includes the battery resistance, similarly the 2 ohm resistor in series with the 24 volt battery may be assumed to include the resistance of the battery. Hence, when the sources are short-circuited, their equivalent resistances remain in the circuit. The resultant circuit is shown in Figure 23b

and it can be seen that the resistance from a to b consists
of a 4 ohm and 2 ohm resistance in parallel and the parallel
combination is in series with a 5 ohm resistance. Therefore
the equivalent series resistor is:

$$R_{ab} = \frac{2 \times 4}{2 + 4} + 5 = 6.33 \text{ ohms}$$

The equivalent Thevenin voltage source therefore consists of
a 20 volt source with 6.33 ohm resistor in series as indicated
in Figure 23c

Figure 23c

The advantage of this equivalent circuit will be best appreci-
ated if we demonstrate its immediate application. Let us suppose that in the original circuit we had terminated the terminals
a and b in a 14 ohm resistance load and it is necessary to know
how many watts are converted into heat as a result of the current
that would flow through the 14 ohm resistance. When we consider the circuit of Figure 23a, directly, with a 14 ohm resistance load connected across the terminals a and b, the current
through the 14 ohm load resistance may be found by loop analysis. However, since we have found that we may represent every-

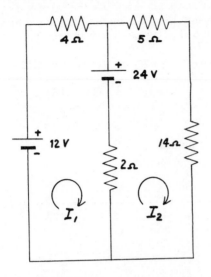

Figure 23d

thing to the left of terminals a-b by an equivalent voltage source as in Figure 23c, the current through the 14 ohm resistance is immediately found from Ohm's law. Thus the current through the load is

$$I = \frac{20}{6.33 + 14} = 0.97 \text{ amperes}$$

The number of watts converted into heat by this current flow in the 14 ohm resistor is

$$P = I^2 R \qquad \text{watts}$$

$$= (0.97)^2 \ 14$$

$$= 13.85 \qquad \text{watts}$$

We shall now check the accuracy of our analysis by verifying that the Thevenin equivalent circuit actually gives us the correct result. Figure 23d represents the circuit of Figure 23a with the 14 ohm resistance across the terminals a-b. To find the current through the 14 ohm resistance we may assume two loop currents are flowing as shown in the figure. We wish to find, of course, I_2.

The two loop equations may be written as follows:

$$6 I_1 - 2 I_2 = -12$$

$$-2 I_1 + 21 I_2 = 24$$

Solving for I_2 we obtain

$$I_2 = \frac{\begin{vmatrix} 6 & -12 \\ -2 & 24 \end{vmatrix}}{\begin{vmatrix} 6 & -2 \\ -2 & 21 \end{vmatrix}} = \frac{120}{122} = 0.97 \text{ amperes}$$

and $P = I_2^2 R$

$$= (0.97)^2 \times 14$$

$$= 13.85 \quad \text{watts}$$

This answer checks with one we obtained from the application of Thevenin's Theorem. We shall see that for all of our simple illustrative examples, the result obtained by Thevenin's Theorem will check with the answer obtained by the general solution for loop currents or node voltages.

The Thevenin Theorem is generally true and can be proven to be true with mathematical rigor. We will accept the truth of the theorem for the present and concentrate on its application as a short-cut for electrical circuit analysis. A method for the application of Thevenin's Theorem seems available from the previous problem. We will stress this method in the problem which follows. A resistive load is being supplied by two generating stations as is shown in Figure 24a

We wish to find the power in the load which will lower the load voltage by 10 per cent. First, we remove the load from the circuit leaving the terminals a and b open-circuited. Some current will now flow from one generator to the other along the line. This current is easily found by the application of Kirchhoff's law to the resulting loop.

$$220 - I(0.10 + 0.05 + 0.05 + 0.10) - 208 = 0$$

or $$I = \frac{12}{0.30} = 40 \text{ amperes}$$

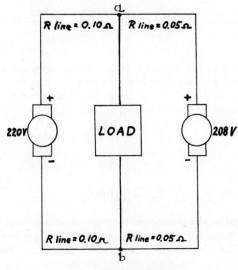

Figure 24a

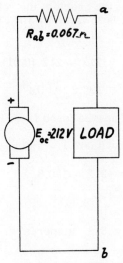

Figure 24b

The voltage drop in the line due to this current flow from the
220 volt generator to the load is

$$40 \times (0.1 + 0.1) = 8 \text{ volts}$$

therefore the open circuit voltage at the load is

$$220 - 8 = 212 \text{ volts}$$

The resistance at the open circuited terminals a-b when the two
generators are replaced by short-circuits is seen to be the
parallel combination of two 0.1 resistors and two 0.05 resistors
in series. Hence the equivalent resistance at the terminals a-b
is

$$R_{ab} = \frac{2 \times (0.1) \times 2 \times (0.05)}{2 \times (0.1) + 2 \times (0.05)}$$

$$= 0.067 \text{ ohms.}$$

Hence the equivalent Thevenin voltage source at the open-
circuited terminals a-b consists of a 212 volt generator in series with a 0.067 ohm resistor as shown in Figure 24b. We now
once more place the load across the terminals a-b in the equivalent circuit of Figure 24b.
From the statement of the problem the voltage at the load, when
it is receiving power, is lowered by 10%. Hence V_{ab}, the voltage
drop across the load is

$$V_{ab} = 212 - 212 (0.01)$$

$$= 190.8 \text{ volts}$$

The load current may now be found from Kirchhoff's equation:

$$212 - 0.067 \text{ I} - 190.8 = 0$$

and

$$I = \frac{212 - 190.8}{0.067}$$

$$= 316 \text{ amperes}$$

The power will then be

$$P = VI \qquad \text{watts}$$

$$= (190.8)\ (316)$$

$$= 60,300 \qquad \text{watts}$$

$$= 60.3 \qquad \text{Kilowatts}$$

This is really a simple solution for a rather difficult problem.
The general procedure for the application of Thevenin's Theorem
consists of the following steps:
1. When the current in a particular load or branch in a circuit
   is desired, remove the load or branch from the circuit leaving the two terminals open-circuited.
2. Measure or compute the open-circuited voltage at these
   terminals.
3. Measure or compute the open-circuited resistance when the
   emfs have been properly removed from the circuit, i.e., the
   emfs have been short-circuited and replaced by their internal resistances.
4. Draw the equivalent circuit consisting of a single series
   circuit comprising the Thevenin generator, E_{oc}, the equivalent resistance R_{ab} and the load resistance R_L.
5. Find the load current, I_L, from the equivalent simple series
   circuit.
These steps are demonstrated in the solution for the load current in the circuit of Figure 25a
Since the load resistance, R_L, is connected across terminals
a-b, we first disconnect R_L from the circuit leaving terminals
a-b open-circuited as in Figure 25b.
We may find the open-circuit voltage a terminals a-b by
measurement or by computation. Here, of course, we compute
the open circuited voltage from Figure 25b. Assume I_1 flows
from terminal c to terminal d through R_1 and R_4 and I_2 flows
from terminal c to terminal d through R_2 and R_3. Since, the
applied voltage is across the terminals c and d,

$$I_1 = \frac{E}{R_1 + R_4}$$

and

$$I_2 = \frac{E}{R_2 + R_3}$$

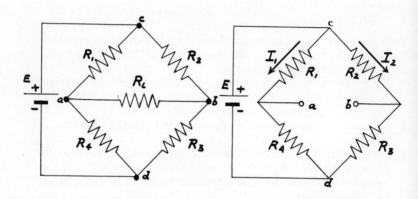

Figure 25a   Figure 25b

The open circuited voltage E_{oc} is equal to the voltage drop from terminal a to terminal b; hence

$$V_{ab} = I_1 R_4 - I_2 R_3$$

or $$V_{ab} = - I_1 R_1 + I_2 R_2$$

Substituting for I_1 and I_2 we obtain

$$V_{ab} = \frac{E}{R_1 + R_4} R_4 - \frac{E}{R_2 + R_3} R_3 \quad \text{from the first equation}$$

$$= E \left\{ \frac{R_4}{R_1 + R_4} - \frac{R_3}{R_2 + R_3} \right\}$$

$$= E \frac{R_4 R_2 - R_1 R_3}{(R_1 + R_4)(R_2 + R_3)}$$

The student should check the second equation for V_{ab} to see
that the same result is obtained.
The resistance, R_{ab}, is now found, after short-circuiting the
source E. The circuit now reduces to that shown in Figure 25c,

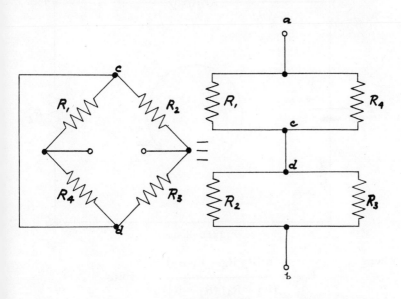

Figure 25c

where it is readily seen that terminals c and d are short-circuited reducing the circuit to parallel combination of R_4 and R_1
in series with the parallel combination of R_3 and R_4.

$$R_{ab} = \frac{R_1\,R_4}{R_1 + R_4} + \frac{R_2\,R_3}{R_2 + R_3}$$

We now draw the equivalent circuit as in Figure 25d

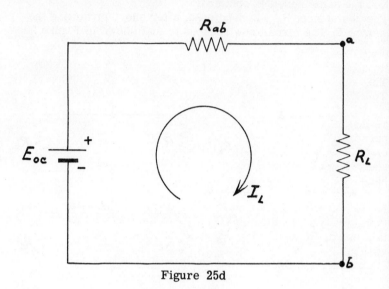

Figure 25d

where

$$E_{oc} = \frac{E (R_2 R_4 - R_1 R_3)}{(R_2 + R_4)(R_1 + R_3)} \quad \text{volts}$$

and

$$R_{ab} = \frac{R_1 R_4}{R_1 + R_4} + \frac{R_2 R_3}{R_2 + R_3} \quad \text{ohms}$$

The load current may now be computed from the simple series circuit and it is

$$I_L = \frac{E_{oc}}{R_{ab} + R_L} \quad \text{amperes}$$

It should be recognized that the solution for the load current by the application of Thevenin's Theorem is really a short-cut in the computation in this example. A straight forward approach by the loop current or node voltage method would necessitate the solution of a third order determinant.

5.5 Norton's Theorem—Current Source Equivalent Circuit

Thevenin's Theorem yields an equivalent voltage source for a two terminal network containing emfs. In a previous chapter we learned how to convert a voltage source and its series internal resistance into an equivalent current source and its internal shunt conductance. The equivalent current source for the Thevenin equivalent voltage source is known as Norton's Theorem. We may state Norton's Theorem as follows:

Any two terminal linear network may be replaced, in so far as the external circuit calculations are concerned, by a simple parallel circuit consisting of a current source, equal to the short circuit current of the network, and a shunt conductance which is equivalent to the conductance of the network when all the emfs have been properly removed.

Consider the circuit of Figure 26a. We shall find both the Norton (current source) equivalent circuit and the Thevenin (voltage source) equivalent circuit at terminals a and b, and obtain the relationship that are valid for this conversion. When we short circuit the terminals a-b we effectively also short-

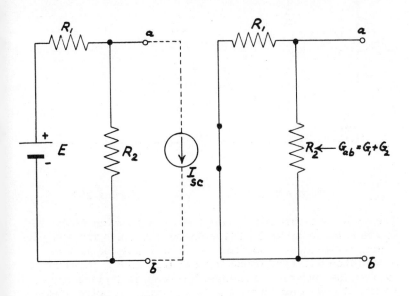

Figure 26a   Figure 26b

circuit R_2, hence the short circuit current, I_{sc}, is seen to be:

$$I_{sc} = \frac{E}{R_1}$$

The equivalent shunt conductance when the emf is short circuited is found from Figure 26b and is seen to be equal to

$G_1 + G_2$ or $\dfrac{R_1 + R_2}{R_1 \, R_2}$ mhos

The Norton current source equivalent circuit is shown in Figure 26c.

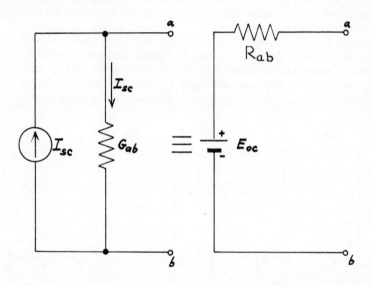

Figure 26c   Figure 26d

The equivalent Thevenin voltage source is shown in Figure 26d. E_{oc} is found from circuit 26a when terminals a-b are open circuited. Under those conditions a current I^1 may be seen to flow in the circuit and producing a voltage drop $I^1 R_2$ which is equal to the open circuit voltage. I^1 flowing in the circuit of Figure 26a when terminals a and b are open-circuited is easily found as

$$E_{oc} = I^1 R_2 = \frac{E}{R_1 + R_2} R_2$$

The resistance, R_{ab}, when the source is short circuited is found to be $R_{ab} = \dfrac{R_1 R_2}{R_1 + R_2}$. The equivalent voltage source comprising E_{oc} and R_{ab} in series must be equal to the equivalent current source. This equivalence is immediately established when we find the open circuit voltage at terminals a-b in the circuit of Figure 26c. Here the open circuit voltage is

$$V_{ab} = \frac{I_{sc}}{G_{ab}} = \frac{E}{R_1} \frac{R_1 R_2}{R_1 + R_2} = E \frac{R_2}{R_1 + R_2} = E_{oc}$$

and since $R_{ab} = \dfrac{1}{G_{ab}}$ the equivalence of the two circuits is established. We shall begin with the current source equivalent circuit and convert it to the Thevenin equivalent voltage source. Therefore, in the circuit of Figure 26c we first find the open circuit voltage. This we found to be

$$E_{oc} = \frac{I_{sc}}{G_{ab}} = I_{sc} R_{ab}$$

Now we open-circuit the current source, I_{sc}, (voltage sources are short-circuited) and compute the R_{ab}, the resistance at terminals of a-b, when the emfs are properly removed. We find here that

$$R_{ab} = \frac{1}{G_{ab}}$$

$$= \frac{R_1 R_2}{R_1 + R_2} \text{ ohms.}$$

The equivalent Thevenin voltage source is therefore identical with the circuit of Figure 26d, which was found to be the equivalent voltage source for the circuit of Figure 26a
Hence the basic relationships between the Norton and Thevenin circuits are

$$E_{oc} = I_{sc} R_{ab}$$

$$R_{ab} = \frac{1}{G_{ab}}$$

We shall now apply Norton's Theorem to demonstrate its use as a short-cut in circuit analysis. In the circuit of Figure 27a we desire to find the current through the 12 ohm resistance load which is being supplied by a current generator and its internal shunt conductance and a voltage source and its internal series resistance.

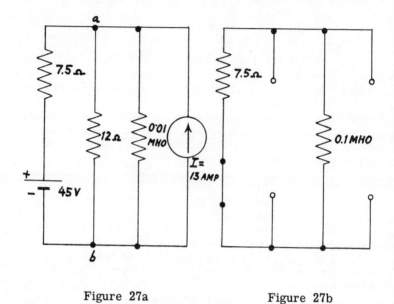

Figure 27a   Figure 27b

First, we remove the load from terminals a-b and short circuit them to find the short circuit current, I_{sc}, for the equivalent Norton current source. When we short circuit terminals a-b we also short circuit the conductance of the current source, hence,

$$I_{sc} = 13 + \frac{45}{7.5}$$

$$= 19.1 \quad amperes$$

The equivalent shunt conductance for the Norton generator is found by properly removing the sources in the circuit and measuring or computing the equivalent conductance at the open circuited terminals a and b. This conductance is most readily

found when we redraw the circuit to remove the emfs properly
from the circuit as is shown in Figure 27b. It is important to
keep in mind the fact that a voltage source is short circuited
while a current source is open circuited. From the circuit of
Figure 27b we see that the Norton equivalent shunt conductance
is

$$G_{ab} = 0.10 + \frac{1}{7.5} \quad \text{mhos}$$

$$= 0.234 \quad \text{mhos}$$

The equivalent Norton generator, therefore, consists of a current source $I_{sc} = 19.1$ amperes and an equivalent shunt conductance $= 0.234$ mhos terminated in a-b. We now connect
the load resistance of 12 ohm to terminals a and b and compute
the load current from the equivalent circuit of Figure 27c

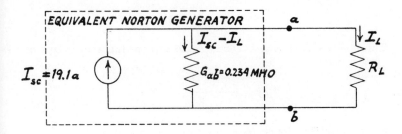

Figure 27c

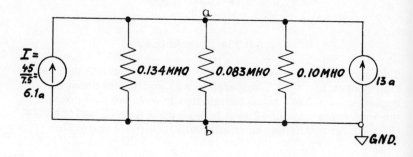

Figure 27d

The load current, I_L, is found readily from the application of Ohm's law:

$$V_{ab} = \frac{I_{sc} - I_L}{G_{ab}} \quad \text{volts}$$

but V_{ab} is also found from the following equation

$$V_{ab} = I_L \, R_L \quad \text{volts}$$

Therefore

$$\frac{I_{sc} - I_L}{G_{ab}} = I_L \, R_L$$

and solving for I_L, the load current we obtain:

$$I_L = \frac{I_{sc}}{1 + R_L \, G_{ab}} \quad \text{amperes}$$

$$I_L = \frac{19.1}{1 + 12\ (0.234)} \quad \text{amperes}$$

$$= 5.02 \quad \text{amperes}$$

We shall check this answer by solving the original circuit of Figure 27a by the node-voltage method. We may simplify the circuit by converting the voltage source to a current source and write the single node-voltage equation which results. The simplified circuit is shown in Figure 27d.

Since there are only the two nodes a and b, we shall consider node b as the reference node and write the single node equation, at a, which completely defines this circuit.

$$\text{Current leaving node a} = \text{Current entering node a}$$

$$V_a\ (0.134 + 0.083 + 0.10) = 6.1 + 13$$

$$V_a = \frac{19.1}{0.317} = 60.4 \quad \text{volts}$$

Therefore the current through the 12 ohm resistance load is

$$I_L = V_a\ G_L$$

$$= (60.4)\ (0.083)$$

$$= 5.0 \quad \text{amperes}$$

For review we will also use the branch current method to solve for the load current. Starting once more with the circuit of Figure 27a we see that the 12 ohm resistance load is in shunt with the 0.10 mho (= 10 ohm) conductance between the terminals a-b. The resultant resistance between the terminals is

$$R_{ab} = \frac{12 \times 10}{12 + 10} = 5.45 \quad \text{ohms}$$

The simplified circuit is now redrawn in Figure 27e.

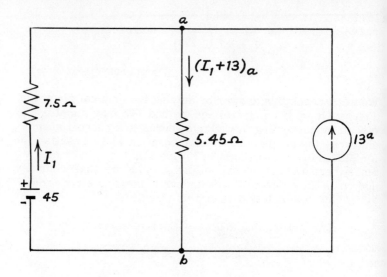

Figure 27e

We will assume a current I_1 flowing in the 7.5 ohm resistance branch in the direction shown in the figure. Applying Kirchhoff's voltage equation we obtain:

$$45 - 7.5\ I_1 - 5.45\ (I_1 + 13) = 0$$

which simplifies to

$$12.95\ I_1 = -70.85 + 45$$

or $\qquad I_1 = -2.0 \qquad$ amperes

This means that the current flowing through the 7.5 ohm resistance is actually in a direction opposite to the one we assumed. It is interesting to check the voltages around this closed loop. Since $I_1 = -2.0$ amperes we have by considering voltage drops in the clockwise direction:

$$-45 + (-2) \times 7.5 + (13 - 2)\ 5.45 = 0$$

$$-60 + 59.95 \cong 0 \quad \text{considering slide rule accuracy.}$$

Since the voltage drop from terminal a to b is 60 volts and the load resistance is between these terminals the load current is readily found to be

$$I_L = \frac{60}{12} = 5 \text{ amperes and checks with the}$$

answer we previously obtained.

5.6 The Maximum Power Theorem

In chapter 1 we saw that the great development in the electrical industry in the past century was due mainly to the ease with which electrical energy can be transmitted over very large distances. The transmission of energy in electrical form covers an extremely wide range, from energy which may be transmitted at a rate of millions of watts over high-voltage alternating current transmission lines at one extreme; to the other extreme where we find that for the communication of intelligence by radio, television, telephone and telegraph the transmission of relatively small amounts of energy is required. Circuits which are used to transmit large amounts of electrical power demand high efficiency and close voltage regulation while many communication circuits are involved only with the transmission of the maximum power that is possible to the load regardless of the efficiency of the system. In this section we will concern ourselves with the transmission of the maximum power possible to the load and in the next section deal with the efficiency of power transmission and voltage regulation.

We have seen that any two terminal network containing linear resistances and emf's may be replaced by an equivalent Thevenin voltage source, representing the open circuited voltage at the two terminals, E_{oc}, and a series resistance, R_{oc}, representing the open circuit resistance at the two terminals when the emf's have been properly removed from the circuit. For any two terminal network we generally would like to know how the terminal voltage differs from the open circuit voltage when power is being delivered by the network to a load placed across the two terminals. Thus, under loaded conditions, when terminal current is flowing we would like to know the volt-ampere characteristics at the terminals. From the equivalent Thevenin circuit for the two terminal network we obtain the relationship between the terminal voltage and the terminal current as

$$V_t = E_{oc} - I_t R_{oc} \quad \text{(volts)} \qquad \text{Eq. 5.81}$$

where V_t is the terminal voltage and I_t is the terminal current.
Figure 28a shows graphically the volt-ampere characteristic,
which is seen to be a straight line whose intercept on the ordinate, the voltage, is obtained by setting $I_t = 0$. Similarly, the
intercept on the abscissa, the current, is found by setting
$V_t = 0$. Therefore when

$$\left. \begin{array}{l} I_t = 0 \\ V_t = E_{oc} \end{array} \right\} \quad \text{and} \quad \begin{array}{l} V_t = 0 \\ I_t = \dfrac{E_{oc}}{R_{oc}} \end{array}$$

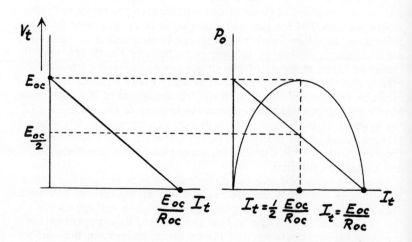

Figure 28a   Figure 28b

The equation 5.81 may be multiplied by the terminal current,
I_t, to obtain the relationship between the output power and the
terminal current.

$$V_t I_t = E_{oc} I_t - I_t^2 R_{oc} \quad \text{(watts)} \qquad \text{Eq. 5.82}$$

But, $V_t I_t$ is equal to the output power, P_O and $E_{oc} I_t$ is equal
to the input power, P_{IN}, while $I_t^2 R_{oc}$ represents the power lost

in the transmission. Thus, equation 5.82 represents the output power as a function of the terminal current.

$$P_O = E_{oc} I_t - I_t^2 R_{oc} \qquad \text{Eq. 5.83}$$

This relation is shown graphically in Figure 28b. The volt-ampere characteristic shown in Figure 28a is also reproduced in Figure 28b. Figure 28b as well as equation 5.83 indicate that P_O, the output power is equal to zero at $I = 0$ and $I = \dfrac{E_{oc}}{R_{oc}}$ where it is seen that $V_t = 0$. The curve also indicates that the slope of the power curve goes through zero at $I_t = \dfrac{1}{2} \dfrac{E_{oc}}{R_{oc}}$ and $V_t = \dfrac{E_{oc}}{2}$ indicating that the power reaches its maximum

value at this point. These results can be established mathematically by the use of the calculus as follows: The expression for the output power, equation 5.83, is differentiated with respect to the current I_t, and the derivative set equal to zero. The terminal current, I_t, may then be solved for the value which yields maximum power. Hence

$$\frac{d P_O}{d I_t} = E_{oc} - 2 I_t R_{oc} = 0$$

whence $\qquad I_t = \dfrac{1}{2} \dfrac{E_{oc}}{R_{oc}} \qquad$ for the maximum value of P_O.

The maximum value of the output power is now found by substituting this value of I_t into equation 5.83

$$(P_O) \max = E_{oc} \cdot \frac{E_{oc}}{2 R_{oc}} - \left(\frac{E_{oc}}{2 R_{oc}} \right)^2 R_{oc}$$

$$= \frac{E_{oc}^2}{4 R_{oc}} \qquad \text{watts}$$

The value of the terminal voltage when the output power is a maximum is found by substituting $I_t = \dfrac{1}{2}\dfrac{E_{oc}}{R_{oc}}$ into the volt-ampere characteristic equation, 5.81.

$$V_t = E_{oc} - I_t R_{oc}$$

$$= E_{oc} - \dfrac{E_{oc}}{2R_{oc}}R_{oc}$$

$$= \dfrac{E_{oc}}{2}$$

It is also very instructive to consider the power lost in transmission when the load is receiving the maximum power. Since the power input is given by $E_{oc} I_t$

$$P_{IN} = E_{oc} I_t$$

$$= E_{oc}\dfrac{E_{oc}}{2 R_{oc}}$$

$$= \dfrac{E_{oc}^2}{2 R_{oc}} \quad \text{when the maximum power}$$

is delivered to the load. Since the maximum power delivered to the load was found to be

$$(P_o)\max = \dfrac{E_{oc}^2}{4 R_{oc}}, \quad \text{therefore the power lost}$$

in transmission will be

$$P_{IN} - P_{out} = \dfrac{E_{oc}^2}{2 R_{oc}} - \dfrac{E_{oc}^2}{4 R_{oc}}$$

$$= \dfrac{E_{oc}^2}{4 R_{oc}}$$

Of course, the power lost in transmission may also be obtained from the equation:

$$P_{loss} = I_t^2 \, R_{oc}$$

$$= \frac{E_{oc}^2}{4 \, R_{oc}^2} \, R_{oc}$$

$$= \frac{E_{oc}^2}{4 \, R_{oc}} \quad \text{watts}.$$

Thus 50% of the power is lost when the maximum power is being delivered to the load.

In communication circuits, it is often necessary to operate under the conditions of maximum power transfer, even though it means operating at 50% efficiency. We, therefore, would like to find the relation between the resistance of the two terminal network and the resistance of the load for maximum power transmission. A two terminal network which is capable of sustaining a constant emf will generally have an internal resistance associated with this energy source. In addition a transmission system usually also contains conductor resistance due to the inter connecting wires between the source and the load. All the resistance associated with the supply and distribution system may be represented by a resistance R_g in series. In Figure 29 we let R_L represent the load resistor which is capable of being adjusted to receive variable power.
The resultant current flowing through the load is

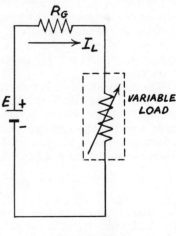

Figure 29

$$I_L = \frac{E}{R_g + R_L}$$

The load power is

$$P_L = I_L^2 \, R_L \qquad \text{watts}.$$

$$= \frac{E^2 \, R_L}{(R_g + R_L)^2}$$

$$= \frac{E^2 \, R_L}{R_g^2 + 2 \, R_g R_L + R_L^2} \text{(watts)}$$

Eq. 5.84

Dividing top and bottom of the right hand side of equation 5.84 by R_g R_L results in the following expression:

$$P_L = \frac{E^2/ R_g}{\dfrac{R_g}{R_L} + \dfrac{R_L}{R_g} + 2} \quad \text{(watts)} \qquad \text{Eq. 5.85}$$

Figure 30 represents a graphical plot of equation 5.85. Several very significant facts can be gained from a careful inspection of this curve. The most striking fact, of course, is that the curve is smooth, starts at zero, as the load resistance varies, rises to a maximum value and then decreases slowly to zero. The abscissa has been chosen as $\dfrac{R_L}{R_g}$, the ratio of the variable load resistance to the fixed internal resistance of the supply and distribution system, first because this ratio is dependent only on the variable load resistance and secondly because we are trying to find the relation between these two quantities for the maximum power transmission. The curve shows us that the output power is zero when $\dfrac{R_L}{R_g}$ is zero and that as $\dfrac{R_L}{R_g}$ is increased, the output increases to a maximum for $\dfrac{R_L}{R_g}$ = 1 and then decreases slowly as $\dfrac{R_L}{R_g}$ increases. Although not shown on the curve, equation 5.85 shows us that as R_L approaches infinity the output power approaches zero. This of course is to be expected because physically when R_L, the variable load resistance, becomes infinite the terminals become open circuited and the load current, I_L, goes to zero so that the output power is zero.

The most significant information that we have obtained from the curve is that the maximum power is delivered to the load, as the load resistance is varied, when the load resistance is made equal to the internal resistance of the generator and distribution system. Stated mathematically, the maximum power is delivered to the load, as the load resistance is varied, when:

$$R_L = R_g \qquad \text{Eq. 5.86}$$

The maximum power that is transmitted is obtained from equation 5.85 when R_L is made equal to R_g.

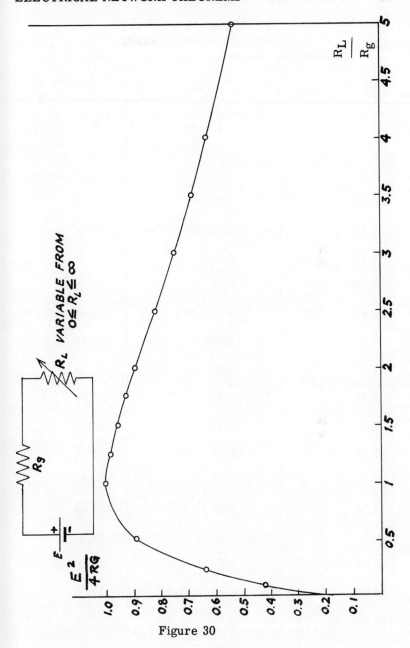

Figure 30

Table 5-1
Power Transfer Theorem
Generator Resistance Fixed—Load Variable

$\dfrac{R_L}{R_g}$	$P = \dfrac{\dfrac{E^2}{R_g}}{2 + \dfrac{R_L}{R_g} + \dfrac{R_g}{R_L}}$	Percent of $P_{MAX} = \dfrac{E^2}{4\,R_g}$
0.00	0.00	0.00 percent
0.10	0.43 $\dfrac{E^2}{4\,R_g}$	43.00 "
0.25	0.64 "	64.00 "
0.50	0.89 "	89.00 "
0.75	0.98 "	98.00 "
1.00	1.00 "	100.00 "
1.25	0.985 "	98.50 "
1.50	0.96 "	96.00 "
1.75	0.93 "	93.00 "
2.00	0.89 "	89.00 "
2.50	0.82 "	82.00 "
3.00	0.75 "	75.00 "
3.50	0.69 "	69.00 "
4.00	0.64 "	64.00 "
5.00	0.54 "	54.00 "
10.00	0.43 "	43.00 "
100.00	0.03 "	3.00 "
∞	0.00 "	0.00 "

Therefore

$$(P_L)_{max} = \frac{E^2}{4 R_g} \quad \text{(watts)} \quad \text{Eq. 5.87}$$

The ordinate of the power curve may therefore be ploted in terms of the ratio of the power P_L, at any value of R_L to the maximum power $(P_L)_{max}$, i.e.

$$\frac{P_L}{(P_L)_{max}} = \frac{4}{\dfrac{R_g}{R_L} + \dfrac{R_L}{R_g} + 2}$$

When this is done the maximum value of the ordinate is unity and it occurs when $R_g = R_L$. Such a curve is referred to as a normalized universal power curve which is generally applicable and not restricted to any particular generator and load. Our curve, which is shown in Figure 30, has the ordinate

plotted so that its maximum value is $\dfrac{E^2}{4 R_g}$ and all other points

indicate the percentage of the maximum value that is transmitted as $\dfrac{R_L}{R_g}$ is varied.

Since the power input to the circuit is given by

$$P_{IN} = E \times I_L \quad \text{(watts)}$$

$$= E \times \frac{E}{R_g + R_L}$$

and when $R_L = R_g$, $\quad P_{IN} = \dfrac{E^2}{2 R_g} \quad \text{(watts)}$

Hence, at maximum power transfer, one half of the available power is lost in transmission. The efficiency of transmission is generally defined as the ratio of the power output to the power input in per cent, therefore, the efficiency is

$$\frac{(P_L)_{max}}{P_{IN}} = \frac{\dfrac{E^2}{4\,R_g}}{\dfrac{E^2}{2\,R_g}} \times 100\%$$

$$= 0.50 \times 100\%$$

$$= 50\%$$

The voltage across the load, V_L, when maximum power is being delivered to the load can be obtained from Ohm's law:

$$V_L = I_L\,R_L \qquad \text{(volts)}$$

$$= \frac{E}{R_g + R_L}\,R_L$$

and when $R_L = R_g$

$$V_L = \frac{E}{2} \text{ volts at maximum power.}$$

The voltage regulation for a transmission system is generally defined as the ratio of difference of the input and output voltage to the output voltage for the rated load. Hence the voltage regulation for maximum power transmission is

$$\frac{V_{IN} - V_L}{V_L} = \frac{E - E/2}{E/2} \times 100\%$$

$$= 100\%$$

We find that for communication systems the maximum power for the transmission of intelligence is more important than the cost of slight power losses. Hence for communication systems it is quite practical to operate at 50 per cent efficiency and 100 per cent voltage regulation.

Instead of plotting the power delivered to the load, as the load resistance is varied, we may obtain the same results by mathematical analysis but we again have to resort to the calculus. Starting with equation 5.84 we differentiate the expression for the power with respect to R_L and set the resulting expression equal to zero. Thus:

$$P_L = \frac{E^2 \, R_L}{(R_g + R_L)^2}$$

$$\frac{d \, P_L}{d \, R_L} = E^2 \, \frac{(R_g + R_L)^2 - 2 \, (R_g + R_L) \, R_L}{(R_g + R_L)^4} = 0$$

which may now be solved for R_L yielding

$$R_L = R_g$$

whence $\qquad (P_L)_{max} = \dfrac{E^2}{4 \, R_g} \quad (watts)$

A numerical example involving maximum power transfer will now be discussed. Consider the circuit shown in Figure 31a which represents a Wheatstone bridge circuit used for the electrical measurement of resistance. The circuit can usually be simplified to four arms which are symmetrically wired into a diamond shape; two terminals are connected to a source of emf while the opposite two are connected together through a galvanometer, an instrument that is extremely sensitive to very minute electrical currents. Two of the arms are fixed resistors, one is a variable standard while the fourth is the unknown resistance whose value is to be determined. The usual operation depends upon the adjustment of the variable standard until the galvanometer shows no deflection. Under these conditions, a definite relationship is established among the four resistances comprising the bridge and since the three arms are of known resistance value, the unknown arm is easily computed. Since the galvanometer is to respond to very minute changes in current it may oscillate, thus take a long time to reach a steady delection. One requirement for the proper operation of a galvanometer is that the resistance of the circuit external to the galvanometer shall be very nearly equal to a value, which is marked by the manufacturer on the galvanometer and referred to as the critical external damping resistance. If the external resistance is much lower than this value, the galvanometer is very sluggish and reaches its steady deflection too slowly, if the external resistance is much larger than the critical value, the galvanometer will oscillate and take a long time to reach a steady deflection. In Figure 31 we wish to determine what critical external damping resistance the galvanometer we choose to use in this bridge circuit will have to have. If it is desired to transfer the maximum power to the galvanometer, what should the galvanometer resistance be? What voltage will appear across the galvanometer terminals?

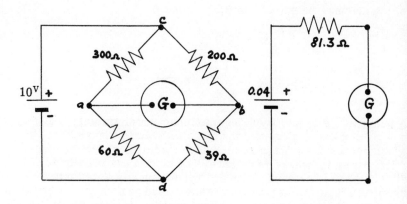

Figure 31a   Figure 31b

First we shall consider the galvanometer terminals open circuited and obtain an equivalent Thevenin voltage source at the terminals a-b. The open circuited voltage is readily obtained since the terminals c-d are directly across the 10 volt source, resulting in a current of 0.028 amperes through the 300 ohm and 60 ohm branches in series and a current of 0.042 amperes through the series 200 ohm and 39 ohm branches. Therefore the voltage drop V_{ab} is given by:

$$V_{ab} = V_{ad} + V_{db}$$

$$= I_{ad} R_{ad} - I_{bd} R_{bd}$$

$$= (0.028) (60) - (0.042) 39$$

$$= 1.68 - 1.64$$

$$= 0.04 \quad \text{volts}$$

Second, we find the equivalent open circuited resistance at the terminals a-b, when the emf is short circuited. When the emf is short circuited the terminals c and d are shorted together, so that looking back from the terminals a-b we see that the 300 ohm and the 60 ohm resistors are in parallel and that this parallel combination is in series with the parallel combination of the 200 ohm and the 39 ohm resistors. Thus the resistance from terminal a to b is

$$R_{ab} = \frac{300 \times 60}{300 + 60} + \frac{200 \times 39}{200 \times 39}$$

$$= 50 + 31.3$$

$$= 81.3 \quad \text{ohms}$$

Therefore, the circuit external to the galvanometer appears as the equivalent Thevenin circuit shown in Figure 31b.
Thus, a galvanometer whose external critical damping resistance is very nearly 80 ohms should be selected for proper operation.
From the equivalent circuit we see that for maximum power transfer to the galvanometer, the galvanometer resistance should be equal to 81.3 ohms. Under these conditions the voltage across the galvanometer terminals will be 0.02 volts and the maximum power transferred will be

$$P_{max} = \frac{(0.02)^2}{4(81.3)} = 1.23 \times 10^{-6} \quad \text{watts}$$

$$P_{max} = 1.23 \quad \text{micro watts}$$

The current through the galvonometer will be

$$I = \frac{P_{max}}{V_{ab}} = \frac{1.23 \times 10^{-6}}{0.02}$$

$$= 61.5 \times 10^{-6} \quad \text{amperes}$$

$$= 61.5 \quad \text{micro amperes}$$

If it is desired that the current through the galvanometer should be zero, then the voltage across the galvanometer terminals should also be zero. Referring to figure 31a, if R_{ac} and R_{bc} are fixed resistors and R_{bd} is the adjustable standard while R_{ad} is the unknown resistance to be determined, then for balancing the bridge we adjust the resistance R_{bd} until zero current flows in the galvanometer resulting in a zero potential difference across a-b, i.e.,

$$V_{ab} = 0$$

and since

$$V_{ab} = V_{ad} + V_{db} = 0$$

$$0 = I_{ad} R_{ad} - I_{bd} R_{bd}$$

or

$$I_{ad} R_{ad} = I_{bd} R_{bd}$$

$$V_{ab} = V_{ac} + V_{cb} = 0$$

$$0 = - I_{ca} R_{ca} + I_{cb} R_{cb}$$

or

$$I_{ca} R_{ca} = I_{cb} R_{cb}$$

but since terminals a and b are at the same potential and the galvanometer current is zero

$$I_{ca} = I_{ad}$$

and

$$I_{cd} = I_{bd}$$

Then

$$\frac{I_{ca} R_{ca}}{I_{ad} R_{ad}} = \frac{I_{cb} R_{cb}}{I_{bd} R_{bd}}$$

reduces to

$$\frac{I_{ca} R_{ca}}{I_{ca} R_{ad}} = \frac{I_{cb} R_{cb}}{I_{cb} R_{bd}}$$

which simplifies to

$$\frac{R_{ca}}{R_{ad}} = \frac{R_{cb}}{R_{bd}}$$

and R_{ad}, the unknown resistances is measured by

$$R_{ad} = \frac{R_{ca}}{R_{cb}} R_{bd}$$

hence in our circuit if R_{bd} is adjusted from 39 volts to 40 volts, then

$$V_{ab} = 0 \quad \text{and the bridge is in balance, i.e.,}$$

the galvanometer current is zero.

$$R_{ad} = \frac{300}{200} \times 40$$

$$= 60 \quad \text{ohms.}$$

CHAPTER 6—THE TRANSMISSION OF ELECTRICAL POWER

6.1 Efficiency and Voltage Regulation

In the previous chapter we have seen that for the communication of intelligence it is quite practical to operate at low efficiency and high voltage regulation. Such transmission for power systems, involving large amounts of power is extremely impractical. We have also seen that whenever an electric current flows some energy is always transformed into heat. Thus since $I^2 R$ loss is always present in the conductors associated with the source and the transmission lines, whenever electrical energy is being delivered, the most efficient system of transmission is always less than 100 per cent. Direct current transmission is generally limited to low voltage transmission of the order of 600 volts and under 5000 volts for short distance transmission. For such low voltage systems the $I^2 R$ loss in the conductors of the transmission lines is practically the only loss. We shall define the efficiency of transmission as the ratio of the useful output power to the total input power to the transmission line. Thus the heating of the conductors of the transmission lines due to their resistance represents energy which is not available to the load and hence is wasted energy since $I^2 R$ conversion is not reversible. Moreover since the conductors generally require some sort of insulation the heating effect when current flows may damage the insulation as well as the conductor if it is excessive. Of course excessive heating is usually due to excessive current flowing through a conductor. Certain safety precautions are specified by the national Board of Fire Underwriters in a National Electrical Code. The National Electrical Code specifications set the allowable current ratings of copper conductors encased in various types of insulation under very definite conditions of enclosure and ambient temperatures. In a subsequent section we shall examine the resistance of conductors in greater detail and we will be in a better position to understand and use the listings of the National Electrical Code.

In the transmission of large amounts of power it is therefore of prime consideration to design the distribution system to meet the specification regarding the proper conductor current carrying capacity. But, this requirement alone does not guarrantee satisfactory overall performance of the transmission system. It is important economically to maintain as high an efficiency of transmission as possible. We have already defined the efficiency of transmission as the ratio of the useful output power to the total input power to the transmission line. We shall first consider a two wire transmission system, which is shown in Figure 32. We will assume that the total conductor resistance, which is actually distributed throughout the entire length of the two conductors, which form the transmission lines, is equally divided by the two conductors and assign the symbol, $R\ell$, to this resistance as in Figure 32.

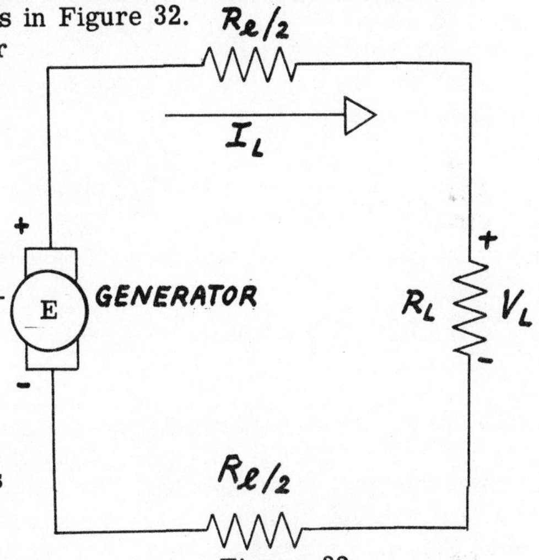

The voltage equation for this simple distribution system is

$$E = V_L + I_L R_\ell$$

The power distribution may be found by multiplying the voltage equation by the current I_L.

$$E I_L = V_L I_L + I_L^2 R_\ell$$

The input power to the transmission system is

$$P_{IN} = E I_L$$

Figure 32

The useful output power is

$$P_o = V_L I_L$$

and thus $I_L^2 R_\ell$ represents the power loss in the transmission.

The efficiency of the system, usually indicated by the Greek letter eta, η is

$$= \frac{P_o}{P_{IN}} = \frac{V_L I_L}{E_L I_L}$$

$$= \frac{V_L}{E} \quad \text{for this simple two wire}$$

system.

When the load on a transmission line is varied from full load to no load there results a potential rise at the receiving end of the transmission line. A measure of this rise in potential is generally referred to as the regulation of the transmission system. The voltage regulation must be kept low for the efficient distribution of the power. We generally define the voltage regulation as the rise in the terminal potential when the load is changed from full load to no load divided by the terminal potential at full load. This definition is quite general and may be used for a generator and an alternating current transformer as well as for a transmission system. When the voltage regulation of a transmission system is low the load is not subjected to undesirable voltage fluctuations.

For our two wire transmission system of Figure 32 the voltage regulation is

$$\text{Voltage regulation} = \frac{(V_L)_{\text{no load}} - (V_L)_{\text{full load}}}{(V_L)_{\text{full load}}}$$

$$\text{Voltage regulation} = \frac{E - V_L}{V_L}$$

Both the efficiency and regulation are generally expressed in per cent. The National Electrical Code specifies that voltage regulation equal to or less than one per cent is adequate for constant voltage lighting systems while for power systems other than lighting a maximum of three per cent is considered adequate.

Since voltage regulation is a measure of the voltage fluctuation at the receiving end of a transmission line and the efficient operation of any electrical load which may be connected to the end of the line is dependent on the voltage at which it operates, it becomes desirable, for efficient operation, to maintain the voltage regulation within the small limits specified above. In primary transmission lines, that is, those feeding power substations, larger values of voltage regulation are possible for two reasons; first because no large load variations are expected and second, voltage fluctuations may be partially compensated in the secondary distribution system.

In Figure 32 the transmission line is 6 miles long and is made of conductor whose resistance is 0.03 ohm per mile. What will be the efficiency of the two wire system when the generator delivers 50 kilowatts at 500 volts.

The current supplied by the generator may be found from

$$I = \frac{P}{V}$$

$$I = \frac{50,000 \text{ watts}}{500 \text{ volts}} = 100 \text{ amperes}$$

The total line resistance is

$$R_\ell = 2 \times 6 \times 0.03$$

$$= 0.36 \text{ ohms}$$

Therefore the power loss in the line is

$$I^2 R_\ell = (100)^2 (0.36)$$

$$= 3.6 \text{ kilowatts}$$

Hence the useful output power is

$$P_o = 50.0 - 3.6$$

$$= 46.6 \text{ kilowatts}$$

The efficiency is

$$\eta = \frac{46.4}{50.0} \times 100$$

$$= 92.8\%$$

We may, of course, also express efficiency as

$$\eta = \frac{P_{IN} - P_{loss}}{P_{IN}} \times 100\%$$

where P_{IN} =   the power input
and P_{loss} = the $I^2 R_\ell$ heat losses

Hence
$$\eta = \frac{E I - I^2 R_\ell}{E I} \times 100\%$$

$$= \left(1 - \frac{I R_\ell}{E}\right) \times 100\%$$

This may also be expressed as

$$\eta = \left(1 - \frac{E I R_\ell}{E^2}\right) \times 100\%$$

which reduces to

$$\eta = \left(1 - \frac{P_{IN} R_\ell}{E^2}\right) \times 100\%$$

Thus for the same conductor resistance and power supplied by the source, the efficiency increases with an increase in the voltage. Hence, if the generator were to supply 50 kilowatts at 750 volts the efficiency would increase to 96.8 per cent, i.e.

$$\eta = \left(1 - \frac{50,000 \times 0.36}{(750)^2}\right) \times 100\%$$

$$= 96.8\%$$

This of course assmes that the losses are only those due to resistance loss in the conductors, but this is substantially true for voltages not exceeding 5000 volts. Hence, within this voltage range the voltage should be made as high as it is conveniently possible for efficient and economical transmission. The voltage regulation, when the generator is delivering 50 kilowatts at 500 volts is found by noting that when the useful output power is 46.4 kilowatts the load current is 100 amperes. Hence the load voltage is

$$V_L = \frac{P_L}{I_L}$$

$$= \frac{46,400}{100} = 464 \text{ volts at full load. At no load}$$

the load current is zero so that the load voltage is equal to the supply voltage, i.e., 500 volts.

Therefore, the voltage regulation is given by

$$\text{Voltage regulation} = \frac{500 - 464}{500} \times 100\%$$

$$= 7.4\%$$

When the supply voltage is increased to 750 volts while the generator is still delivering 50 kilowatts, the full load or rated current is now 66.67 amperes and the useful output power is

$$P_L = 50,000 - (66.67)^2 \, (.36)$$

$$= 50,000 - 1600$$

$$= 48.4 \text{ kilowatts}$$

Therefore, the voltage at full load is

$$V_L = \frac{48,400}{66.67}$$

$$= 726 \quad \text{volts and since at no load, there is no}$$

load current and therefore no voltage drop in the line, the no load receiving end voltage is equal to the supply voltage, i.e., 750 volts. The resulting voltage regulation is

$$\text{Voltage regulation} = \frac{750 - 726}{750} \times 100 \text{ per cent}$$

$$= 3.2 \text{ per cent}$$

Therefore when the supply voltage was increased from 500 volts to 750 volts, that is, by 50 per cent the efficiency was increased from 92 to 96.8 per cent and the voltage regulation was improved by lowering it from 7.4 per cent to 3.2 per cent.

6.2 Ratings of Electrical Equipment

Direct current electrical power equipment is generally designed to supply or deliver a specific amount of power at a definite voltage. Non satisfactory operation is usually the result
when these devices are operated at values of voltage, other
than the rated one. Operating at rated voltage and power, of
course, means operation at rated current. A 100 watt lamp,
which has been designed to operate properly at 115 volts, will
give more light at a higher voltage. But operation at a higher
voltage, means operation at a higher current and hence will reduce the life of the lamp. Operating the lamp at a voltage lower
than the rated voltage means of course inadequate light. Devices
like electrical lamps are designed to be operated in parallel
from a fairly constant voltage source. An electrical motor
which has been designed to deliver 5 horse power (5hp) at 115
volts generally overheats when it is operated from a lower voltage supply.

Other electrical devices are designed to operate in series at
a constant current. Vacuum tube filaments are designed for
parallel operation from a voltage supply or series operation
from a constant current supply, the former are given a voltage
rating, the latter a current rating. A 6 ampere arc light will
operate from a 6 ampere current generator; the generator
will deliver 6 amperes to one or to ten arc lights in
series.

Constant current power transmission is not in common
use for two reasons. First, a series arrangement for power distribution is not satisfactory because any break in the series
circuit means the removal from service all of the devices connected to it. Secondly, the voltages from the terminals of some
of the series devices to the grounded terminal of the constant
current generator may become dangerously high. Of course,
the constant current generator would have to be designed to
provide a fairly constant current with increasing voltage as
the load on the generator increases.

Therefore electrical power is generally supplied at constant
voltage. Motors, generators, lamps and household appliances
are designated by a specific voltage rating and a power or current rating. Insulation generally enables the electrical equipment
to withstand many times the rated voltage between its terminals,
while the current flows only through the intended path. The voltage rating is usually dependent upon the intended service. The
power which a piece of electrical apparatus may deliver or
absorb is primarily determined by the current rating. We have

already seen that currents flowing in conductors will heat
them. The temperature of the conductors increase until an
equilibrium is established between the electrical power converted into heat and the heat power lost. The current rating of
electrical equipment is determined by the final temperature
which the conductor is allowed to reach with safety. In most
electrical apparatus, the maximum safe temperature is that
above which the insulation begins to deteriorate. In lamps, flat
irons, heaters, and other devices dependent on the production
of heat, the rated current is a function of the temperature at
which the conductor itself begins to deteriorate.

In motors and generators, windings of insulated wires
are used and the insulation material commonly used will
deteriorate rapidly at high temperatures even though it
does not burn. Although heat is produced throughout the
entire winding only the surface coil can dissipate this
heat to the surroundings, so that the surface temperature
is usually less than the temperature internal to the coil.
Thus, the American Institute of Electrical Engineers has
designated certain "hot spot" temperature limits for the different types of insulation in various services. These limits are
given a definite temperature rise above an assumed ambient
(room temperature) of $40°$ C and may vary from $35°$ C rise for
cotton, silk and paper insulation to $85°$ C for such materials as
mica and asbestos. Electrical machinery is generally rated for
both continuous and intermittant service. The continuous rated
current is so selected that the temperature will remain below
the specified maximum, no matter how long the rated current
may flow. For intermittent service, electrical machinery may
be operated for short periods of time at substantially higher currents than the rated one, during which time the temperature
does not exceed the maximum specified. Thus the ratings of
electrical power equipment are determined by the allowable
temperature rises, which in turn are fixed by the heat losses in
the equipment. The ratings are usually stated with respect to
voltage and power or voltage and current but these current ratings are controlled to prevent excessive heating which may
result in damage to the equipment. Thus heating is the principal limitation on the power capacity of many electrical devices.
On the other hand, in many types of communication equipment
the current flowing is so small that heating becomes a negligible
factor.

6.3 Three Wire Transmission System

In power systems the cost of the power involved is quite appreciable and it is extremely desirable to keep the power losses as low as possible. Figure 28b shows us how the power varies with the load current and from this curve it is seen that power equipment should be designed for operation as near to the origin of the curve as it is practical. We may further reduce the $I^2 R$ losses in transmission of electrical power by keeping the current, I, as small as possible in transmission. Therefore, if we desire to transmit a fixed amount of power, and maintain the current as low as possible, we must increase the voltage of the transmission. Thus, when we transmit 50,000 watts we may do so in the form of 500 volts at 100 amperes or 250 volts at 200 amperes. But if we transmit the 50,000 watts in the form of 1000 volts at 50 amperes the loss in the transmission line will be $\dfrac{(50)^2}{(100)^2}$ or $\dfrac{1}{4}$ as much as if it were transmitted over the same line at 500 volts at 100 amperes. That is, doubling the voltage at which the power is being transmitted decreases the power loss in the line by 75 per cent.

We may generalize these considerations as follows:

For a fixed amount of power that is being transmitted over a a line, whose conductor resistance does not change, the power loss ($I^2 R$) in the line varies directly with the square of the line current. Since the amount of power is fixed, the line current varies inversely with the voltage of the line. Therefore, the power loss in the line varies inversely with the square of the voltage on the line.

For alternating current transmission, alternating current power can be readily transformed from one voltage level to another by reliable, stationary equipment, known as alternating current transformers, operating at very high efficiencies. The use of alternating current transformers makes possible the transmission of very large amounts of power over very large distances. As an example, large amounts of power may be transmitted for 300 miles at a voltage as high as 300,000 volts. At the point where the power is consumed the voltage may be transformed to a safe and convenient value say 115 volts. Usually between the generator and the consumer, the voltage is transformed more than once. The only limitation with reference to the distance over which the transmission takes place economically will depend upon the voltage being used and the power which is being carried.

For direct current power transmission, higher efficiency of
higher voltage transmission is usually limited by the fact that
the equipment that is needed at the receiving end to reduce the
voltage for normal operation of most electrical equipment
(115 volts) is complicated and expensive. Of course, this
limitation does not exist if the power can be used at the same
voltage at which it is being transmitted. Hence direct current
transmission is limited to comparatively short distances. Direct current railway practice however, when voltages over
3000 volts are used does employ many miles of track from a
single station.

Direct current power is transmitted at the same voltage at
which it is to be used. However, three wire transmission offers
many advantages over two wire transmission. Three wire
transmission usually makes use of a neutral wire which is
maintained at ground or zero potential. This enables transmission to take place at generally double the voltage at which the
power is to be used with the efficiency of the higher voltage
transmission. In three wire transmission the loads are generally
placed between each outside wire and the neutral. The set up for
maximum efficiency is to have the loads on both sides of the
neutral wire "balanced," when this is done no current flows in
the neutral wire. Two generators, in series may be used for
double voltage transmission with a three wire system. Usually
a special three wire generator or a large generator and balancer
set is used with a three wire system.

Figure 33 represents a three wire transmission system. Two
115 volt generators are used to supply a three wire system
which is feeding a 230 volt motor, marked M, and 4 sets of
lamp banks. Thus 230 volts is impressed across the two outside
conductors A and A' and only 115 volts exists between either
one of these wires and the neutral wire N, which is at ground
potential. Therefore the lamps have practically 115 volts across
them while the motor has very nearly the 230 volts it requires.
Moreover, there is no point in the circuit which is more than
115 volts will respect to the ground. This system combines the
advantages of both 115 volts and 230 volts transmission with
practically the efficiency of a 230 volt line transmission system.

6.4 Illustrative Example

A distribution system comprises two generators which are
supplying power to two loads. The first load, I_1, consumes 40
amperes and the second load, I_2, consumes 95 amperes. Both

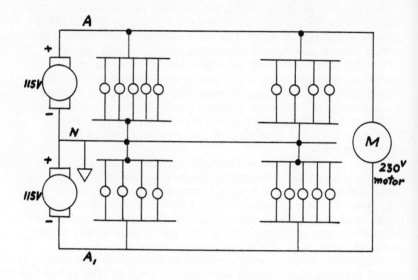

Figure 33

generators are regulated in order to maintain constant terminal potentials, E_1 = 120 volts and E_2 = 128 volts. The resistance of the transmission line is 0.04 ohms for each section of line from genera tor to load and from load to load. We shall determine the overall efficiency of the distribution system, the voltage regulation across the two loads, and the potential required for generator E_1 if the generator loads were made equal. The circuit for this system is shown in Figure 34.

First we shall find the currents I_a and I_b flowing in the directions indicated in Figure 34. Since the current leaving the 128 volt generator is I_a and the current entering the 120 volt generator is I_b then by Kirchhoffs current law

$$I_a = I_b + 135$$

We now may write a Kirchhoff voltage equation about the entire outside loop comprising the generators and transmission lines in a counter clockwise direction starting at the 128 volt generator:

128 - (0.04) 2 I_a - (0.04) 2 (I_a - 95) - (0.04) 2 (I_a - 135) - 120 = 0

which reduces to:

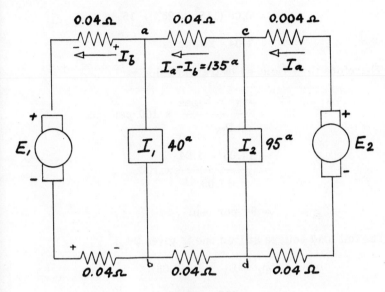

Figure 34

$$2 I_a + I_b = 195$$

but $\qquad\qquad I_a - I_b = 135 \qquad$ and adding the two

equations $\qquad\qquad 3 I_a \qquad = 330$

Therefore $I_a = 110$ amperes and $I_b = -25$ amperes. Since I_b is found to be minus 25 amperes, it means that physically, generator E_1 is supplying 25 amperes to load one while generator E_2 is supplying 15 amperes to load one and 95 amperes to load two. The total input power is readily obtained since

$$P_{IN} = E_1 I_b + E_2 I_a \qquad \text{watts}$$

$$= (120)\,(25) + (128)\,(110)$$

$$= 3000 + 14,080$$

$$= 17.08 \quad \text{kilowatts}$$

The power lost in transmission is given by

$$P_{lost} = (110)^2 (0.08) + (25)^2 (0.08) + (15)^2 (0.08)$$

$$= 0.08 (110^2 + 25^2 + 15^2)$$

$$= 1.04 \quad \text{kilowatts}$$

Therefore the efficiency of transmission is

$$= \frac{P_{IN} - P_{loss}}{P_{IN}} \times 100 \text{ per cent}$$

$$= \frac{17.08 - 1.04}{17.08} \times 100$$

$$= 94 \text{ per cent}$$

The full load voltage at load one is given by

$$V_{ab} = E_1 - I_b (0.08)$$

$$= 120 - 25 (0.08)$$

$$= 118 \text{ volts when 40 amperes are flowing}$$
in load one. The full load voltage at load two is given by

$$V_{cd} = E_2 - I_a (0.08)$$

$$= 128 - 110 (0.08)$$

$$= 119.2 \quad \text{volts when 95 amperes are flow-}$$
ing in load two. The total useful output power is found from

$$P_o = V_{ab} I_1 + V_{cd} I_2$$

$$= (118) (40) + (119.2) (95)$$

$$= 4.72 \text{ kw} + 11.32 \text{ kw}$$

$$= 16.04 \quad \text{kilowatts. This answer may now}$$
be easily checked from the previous computation since the useful output power is also given by:

$$P_O = P_{IN} - P_{loss}$$

$$= 17.08 - 1.04$$

$$= 16.04 \quad \text{kilowatts}$$

To find the voltage regulation it is necessary to find the voltage
at each load when there is no load current. To determine V_{ab}
when the 40 ohm load is removed we redraw the circuit as in
Figure 35a

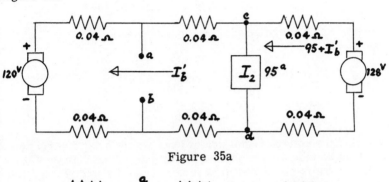

Figure 35a

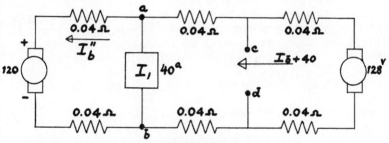

Figure 35b

Assuming that generator E_2 now supplies $95 + I_b'$ amperes as
shown in the figure we may write the following Kirchhoff voltage equation starting at generator E_2 in a counterclockwise
direction:

$$128 - (95 + I_b') (2 \times 0.04) - I_b' (4 \times 0.04) - 120 = 0$$

solving for I_b' we obtain

$$I_b' = 1.67 \quad \text{amperes}$$

and $$V_{ab} = E_1 + I_b' (0.04 \times 2)$$

$$= 120 + 1.667 \times 0.08$$

$$= 120.133 \quad \text{volts at no load}$$

Therefore the voltage regulations for load one is

$$\text{Voltage regulation} = \frac{(V_{ab}) \text{ no load} - (V_{ab}) \text{ full load}}{(V_{ab}) \text{ full load}}$$

or per cent voltage regulation for load one is

$$\text{Voltage regulation} = \frac{120.133 - 118}{118} \times 100 \text{ per cent}$$

$$= 1.8 \quad \text{per cent}$$

To find the voltage regulation at load two we must determine the no load voltage at terminals c and d, that is, the voltage V_{cd} when there is no current flow from terminal c to d. This voltage is readily determined from Figure 35b, where we now assume that generator E_2 is supplying the 40 ampere load as shown in the figure. Once again writing the voltage equation starting with generator E_2 in a counterclockwise direction we obtain:

$$128 - (I_b'' + 40)(4 \times 0.04) - I_b''(2 \times 0.04) - 120 = 0$$

and solving for I_b'' we find that

$I_b'' = 6.67$ amperes. Therefore when the load is disconnected from terminals c and d generator E_2 supplies a total current of 46.67 amperes which includes 40 amperes to the load at the terminals a-b. The no load voltage at terminals c-d may now be found.

$$V_{cd} = 128 - 46.67 (2 \times 0.04)$$

$$= 124.27 \text{ volts with no current flowing}$$

from terminals c to d. The voltage regulation at load two is now determined as:

$$\text{Voltage regulation} = \frac{124.27 - 119.2}{119.2} \times 100 \text{ per cent}$$

$$= 4.2 \text{ per cent}$$

The last question may be answered as follows: In order for the generator loads to be made equal each generator should supply one half of the total load. The total load is

$$I_1 + I_2 = 40 + 95 \text{ amperes}$$

$$= 135 \text{ amperes}$$

Hence each generator should provide 68.5 amperes. A voltage equation may now be written for Figure 34 with both generators providing 68.5 amperes at full load. Starting at generator E_2 in a counterclockwise direction we have:

$$128 - 68.5 (2 \times 0.04) + 26.5 (2 \times 0.04) + 68.5 (2 \times 0.04) - E_1 = 0$$

or
$$E_1 = 128 + 26.5 (0.08)$$

$$= 130 \text{ volts}$$

Thus in order for both generators to carry equal loads, the potential of generator E_1 must be kept at 130 volts.

6.5 The Computation of Resistance

For the proper design and use of electrical equipment it is important to be able to predict the resistance of an electrical conductor from the characteristics of the material from which it is made and from its geometric shape.

Experiment indicates that the resistance of a cylindrical conductor is directly proportional to the length of the conductor. This seems logical since by increasing the length of a conductor we are actually joining in series pieces of conductor, and the resistances in series in any circuit is the sum of its parts. Thus, when we place two identical conducting wires in series the total resistance is doubled; but the total length of conductor is also doubled. Hence, the experimental result, that the resistance of a conductor varies directly with its length seems like a logical conclusion. We also find experimentally that the resistance of a conductor is inversely proportional to its cross-sectional area.

When we place two identical conductors in parallel, the resistance of the combination is one half the resistance of each conductor. We can see that increasing the cross-sectional area of a conductor is equivalent to placing pieces of the conducting material in parallel. Thus we conclude that the resistance of a conductor is directly proportional to its length and inversely proportional to its cross-sectional area. We may express this relation as a mathematical equation.

$$R = \frac{\rho \ell}{A} \text{ ohms} \qquad\qquad \text{Eq. 6.51}$$

R is expressed in ohms when ℓ, the length of the conductor is expressed in meters, A, the cross sectional area is in square meters and ρ (the Greek letter rho) is the proportionality constant, and is given in ohm-meters. Experimentally it is found that the resistance of a conducting material is dependent upon the material. The proportionality constant, ρ, takes this factor into account it is given the name of resistivity or specific resistance and its values for the various materials is listed in published tables. The dimensions of resistivity are resistance times length and therefore depend upon the system of units that are used. In the meter-kilogram-second system, abbreviated MKS, which we use, the dimensions of resistivity are the ohm-meter. This unit is so inconvenient in size that it is rarely used. Instead, resistivity of conductors is measured in microhm-centimeters or ohms-circular mils per foot. The microhm-centimeter finds general use in the laboratory and in scientific investigations, and it is well adapted for calculations involving conductors of square or rectangular cross-section. The ohm-circular mil per foot is much used in industry and very convenient for calculations of round conductors. A sample of a particular conductor is selected and its resistance is found from measurement. The value of resistivity may then be found from equation 6.51. Resistivity is also found from experiment to be a function of temperature. The resistivity of ordinary annealed-copper wire, at 20°C, is found from careful tests which are performed on many samples of the present day grades, as 1.72 microhm-centimeter, i.e. 0.00000172 ohm-centimeter.

Example

The resistance of an annealed copper bar which is 50 inches long and whose cross-sectional area is 0.5 square inches is found from equation 6.51.

50 inches = 127 centimeters
0.5 square inches = 3.23 square centimeters

$$R = \frac{1.72 \times 127}{3.23} \text{ microhms}$$

$$= 67.8 \text{ microhms}$$

$$= 0.0000678 \text{ ohms.}$$

For cylindrical conductors the calculations of resistance are greatly simplified by measuring the cross-sectional area in circular mils. A unit of area may be expressed as a circular mil. We choose this unit as a circle whose diameter is one mil, equal to 0.001 inch. Just as we take a square, one inch on each side, and divide this unit into an area to find the total number of square inches, so we take a cylindrical conductor of circular cross-section and divide it by a circle whose diameter is one mil to see how many times this unit is contained in the circular conductor in circular mils.

A cylindrical conductor has a circular cross-section of diameter d inches. Hence, the area of this conductor, A, is

$$A = \frac{\pi d^2}{4} \text{ square inches}$$

Our measuring unit has a diameter of 0.001 inches. Therefore, the unit area, A_u, is

$$A_u = \frac{\pi (0.001)^2}{4} \text{ square inches}$$

To find out how many times the unit is contained in the area A we form the ratio

$$\frac{A}{A_u} = \frac{\dfrac{\pi d^2}{4}}{\dfrac{\pi (0.001)^2}{4}} = \frac{d^2}{(0.001)^2} = (1000 \, d)^2$$

This means that for any cylindrical conductor of circular cross-section if we want to find the area in circular mils we merely multiply the diameter, in inches, by 1000 and square the product. Resistivity is sometimes considered as being numerically equal to the resistance of a section of a conducting material having a unit length and a unit cross-sectional area. Thus the resistivity in ohm-circular mils per foot is numerically equal to the resistance of a section of conductor having a cross-sectional area of one circular mil and a length of one foot. Such a section of conductor is sometimes called a mil-foot of conductor. What is really meant by the mil-foot is that the resistivity of such a section of conductor is expressed in ohm-circular mils per foot.

Therefore we may restate equation 6.51 in the following form:

$$R = \rho \, \frac{\ell}{d^2} \quad \text{ohms} \qquad\qquad \text{Eq. 6.52}$$

where ρ = resistance in ohm-circular mils per foot

ℓ = length of conductor in feet

d = diameter of the conductor in mils

Thus in our previous example we found that a cylindrical annealed copper bar 50 inches long and having a cross-sectional area of 0.5 square inches had a resistance of 67.8 microhms. From this result we may obtain the resistivity of annealed copper at 20°C in ohm-circular mils per foot.

Since
$$\rho = \frac{R\,A}{\ell} \quad \text{ohm-circular mils per foot}$$

$$\rho = 67.8 \times 10^{-6} \times \frac{\dfrac{4\,(0.5)}{\pi\,(0.001)^2}}{\dfrac{50}{12}}$$

$$= 10.42 \quad \text{ohms}$$

The common expression is, therefore, that the resistance per mil-foot of annealed-copper is 10.4 ohms at 20°C.

We would like to find, now, the resistance of a piece of wire
drawn from the same material but having a diameter of 0.1 inch
and 1 mile long. Since

$$R = \rho \frac{\ell}{d^2}$$

$$= \frac{10.4 \times 5280}{(0.1 \times 1000)^2}$$

then $R = 5.5$ ohms

Copper, aluminum and steel are the most commonly used materials for conductors; copper for its low resistivity, aluminum
for its light weight and steel for its strength. The resistivity of
soft-drawn copper is 1.724 microhm-centimeter and is the metal
in greatest use. Hard-drawn copper has a resistivity of less than
3 per cent greater, 1.772 microhm centimeter, but its tensil
strength is 50 per cent greater and hence it finds extensive use
in transmission lines.

The resistivity of steel is a function of its composition and
treatment but it is always much greater than copper. The
resistivity of steel is approximately 21.6 microhm centimeter
(about 12 times greater than copper) for steel wires. The resistivity for steel rails varies from 13.8 to 21.6 microhm-centimeters.

Aluminum has a resistivity of 2.828 microhm centimeters at
20°C. This is more than one and half times that of copper but
the small weight per cubic centimeter of aluminum more than
makes up for its resistivity. For the same length and weight a
conductor of aluminum has less resistance than copper.

Alloys of metals, generally, have higher resistivities than
any of the constituent metals, usually higher than that of the constituent metal of lowest resistivity. Alloys of copper nickel, zinc,
manganese and chromium are used as resistors. Some alloy resistivities are almost as high as 100 microhm-centimeters.
Table 6-1 at the end of chapter 6 lists some of the materials in
common use in the electrical industry.

It is customary when specifying the grade of the material to
be used as a conductor to specify the conductivity rather than the
resistivity. We have defined conductance as the reciprocal of the
resistance. Hence we designate the reciprocal of the resistivity
as the conductivity. We can also arrive at the same conclusion
from the following consideration. Since conductance is the

reciprocal of the resistance than any one of the factors which increases resistance will decrease the conductance. Thus the conductance is directly proportional to the cross-sectional area and inversely proportional to the length of a conductor. Introducing a proportionality constant which is dependent upon the conducting material we may write:

$$G = \frac{\gamma A}{\ell} \text{ mhos} \qquad\qquad \text{Eq. 6.53}$$

where γ (the Greek letter gamma) is the proportionality constant and in the MKS system of units, its dimensions are mhos per meter when the length is in meters and the area in square meters. Since

$$G = \frac{1}{R} \text{ mhos}$$

and
$$R = \rho\frac{\ell}{A} \text{ ohms}$$

then
$$\gamma = \frac{1}{\rho} \text{ mho per meter}$$

It is customary to arbitrarily select a certain grade of copper as a standard. This standard is then said to have 100 per cent conductivity and all other metals are then rated as a certain per cent conductivity of that standard.

Standard Annealed Copper at 20° C has a resistivity of 1.724 microhm centimeters and a density of 8.89 grams per cubic centi meter. The conductivity of this standard is called 100 per cent. Percentage conductivity of any material is rated as a certain percentage of this standard. The conductivity of Standard Annealed Copper at 20° C is

$$\gamma = \frac{1}{\rho}$$

$$= \frac{1}{0.000001724}$$

$$= 580,000 \quad \text{mhos per centimeter}$$

$$\text{Percentage conductivity} = \frac{\gamma \text{ of material}}{\gamma \text{ of standard annealed copper}} \times 100$$

$$= \frac{\gamma \text{ of material}}{580,000} \times 100$$

$$= \frac{\rho \text{ of standard annealed copper}}{\rho \text{ of material}} \times 100$$

Therefore, if a material has a conductivity of 90 per cent, it means that the material has a conductivity which is 90 per cent of that of standard copper, i.e. 0.90 x 580,000 mhos per centimeter conductivity. This material has therefore a resistivity of $\frac{1.724}{0.90}$ microhm-centimeter. Aluminum has an average conductivity of 61 per cent. Copper generally varies from 98 to 100 per cent. it is possible to obtain copper which is purer than the standard and which will have a conductivity which is greater than 100 per cent. Steel wire has about 10 per cent conductivity.

The presence of a small amount of impurity in a metal greatly reduces its conductivity. Alloys of good conducting materials have very poor conductivities. Thus Nichrome, an alloy of copper, nickel and iron, all good conductors has a conductivity of only 1.5 per cent. Since it is often desirable to get a lot of resistance in a relatively short length of wire, such as in heating elements, alloys like Nichrome become of great practical importance.

We will consider a few illustrative examples:
It is desired to find the resistance per "mil-foot" at 20°C of a particular stock of copper which has a conductivity of 97 per cent. Here 97 per cent =

$$\text{cent. Here 97 per cent} = \frac{\rho \text{ standard annealed copper}}{\rho \text{ copper sample}} \times 100$$

Therefore:

$$\rho \text{ of copper sample} = \frac{10.4 \text{ ohm-circular mils per foot}}{0.97}$$

$$\rho = 10.7 \text{ ohm-circular mils per foot}$$

A solid round aluminum wire, 0.365 inches in diameter has a resistance of 0.672 per mile at 20°C. What is its per cent conductivity. Since resistance is given by

$$R = \rho \frac{\ell}{d^2} \quad \text{ohms}$$

then

$$\rho = \frac{R \, d^2}{\ell} \quad \text{ohms-circular mils per foot}$$

and since 0.365 inches = 365 mils

$$\rho = \frac{0.672 \, (365)^2}{5280}$$

$$= 17.1$$

Per cent conductivity $= \dfrac{\rho \text{ of standard}}{\rho \text{ of material}} \times 100$

$$= \frac{10.4}{17.1} \times 100$$

$$= 61.3 \text{ per cent}$$

If we desire to use a copper conductor, having 96 per cent conductivity and the same resistance per mile as the aluminum wire what should its diameter be?
The resistivity of the 96 per cent copper is

$$\rho = \frac{10.4}{0.96} \quad \text{ohm-circular mils per foot}$$

$$= 10.82$$

and from

$$R = \rho \frac{\ell}{d^2} \quad \text{ohms}$$

we may solve for

$$d^2 = \rho \frac{\ell}{R} \text{ circular mils}$$

hence

$$d^2 = \frac{10.82 \times 5280}{0.672}$$

and

$$d = 291 \quad \text{mils}$$

$$= 0.291 \quad \text{inches}$$

Experimental evidence is available to show that the resistance of any pure metal changes with temperature. Therefore, when specifying the resistivity of a conductor, the temperature is always given, i.e., the resistivity of copper is 10.4 ohm-circular mils per foot at 20° C, while that for aluminum is 17.1. For nearly all pure metals, the temperature resistance relation is practically linear in the temperature interval from -50° C to + 200°C. Thus a graph of the temperature-resistance characteristic is practically a straight line over this limited temperature range. Fortunately, for most practical problems and most conductors, this is usually the operating temperature range. Of course, this does not apply to those conductors which are used as heating elements. Therefore, in calculating the changes of the resistance of conductors with temperature we may make use of the straight line relation. For copper it is found that for each degree that the temperature rises above 20°C, up to about 200° C, the resistance increases 0.393 of one per cent of the value at 20°C. Also, for each degree that the temperature of the copper falls below 20°C, down to about - 50°C, the resistance decreases 0.393 of one per cent of its value at 20°C. The percentage change in resistance, 0.393 of one per cent, is referred to as the Temperature Coefficient of Resistance and for most pure metals this coefficient has almost the same value. Temperature Coefficients of Resistance for metals commonly used in the electrical industry are given in Table 6-1 at the end of the chapter.

The resistivity of copper has been given as 10.4 ohm-circular mils per foot at 20°C hence all computations of resistance of conductors using this value will give the resistance at 20°C. Hence, if we wish to find the resistance of a conductor at any other temperature, we must find the increase or decrease in resistance and add it or subtract it from the resistance at 20°C. The curve showing the relation between the resistance and the temperature of Standard Annealed Copper is shown in Figure 36 The curves for other pure metals would have the intercepts on the temperature axis somewhat different from   -234.5°C and

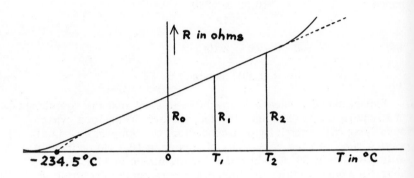

Figure 36

the slope would vary somewhat. An equation may be written for the straight line portion of this curve.

$$\frac{R_2}{R_1} = \frac{234.5 + T_2}{234.5 + T_1}$$

Thus if the resistance of the conductor at one temperature is known, it can be computed for any other temperature, provided that both temperatures lie within the interval for which the relationship is linear. The slope of the curve is

$$M = \frac{R_2 - R_1}{T_2 - T_1}$$

and physically represents the change in resistance per degree change in temperature. If we divide this slope by the resistance at any particular temperature we obtain the Temperature Coefficient of Resistance, α, (the Greek letter alpha). Thus

$$\alpha_1 = \dfrac{\dfrac{R_2 - R_1}{T_2 - T_1}}{R_1} \quad \text{change of resistance per}$$

degree change of temperature per ohm. Since α_1 is dependent upon the resistance R_1, and R_1 depends upon the temperature, then α_1 depends upon the temperature, i.e., the Temperature Coefficient is different for every temperature. The equation involving α_1 may be written in the following form

$$R_2 = R_1 \left[1 + \alpha_1 (T_2 - T_1) \right] \qquad \text{Eq. 6.54}$$

For copper, α_1; the Temperature Coefficient of Resistance is given in Table 6-1 as 0.00393 at 20°C. We may, therefore, find α_2, for any other temperature, T_2 from the equation

$$\alpha_2 = \dfrac{\alpha_1}{1 + \alpha_1 (T_2 - T_1)} \qquad \text{Eq. 6.55}$$

Electrical Machinery is usually guaranteed not to have the temperature of the wires in the coils rise above a given number of degrees when operating under a given load for a given time. The average temperature rise may be found by first measuring the resistance of the coils at room temperatute, about 20°C, then at the end of the run and applying equation 6.54.

The primary coils of a transformer are measured at 20°C and found to be 6.24 ohms. After operating the transformer at full load for 3 hours the resistance is now measured as 7.50 ohms. We wish to find the temperature rise in the coils.

The resistance increase is 7.50 - 6.24 = 1.26 ohms

The percentage increase of resistance is $\dfrac{1.26}{6.24}$ x 100 equals

22.5 per cent. The percentage increase per degree rise above 20°C is 0.393 per cent. Therefore, the rise must be

$$\dfrac{22.5}{0.393} = 51.5°C \text{ to produce a 22.5 per cent}$$

increase in resistance. The equation 6.54 was derived from similar considerations, hence we may apply it directly

$$\frac{R_2}{R_1} = \frac{234.5 + T_2}{234.5 + T_1}$$

or

$$T_2 = \frac{R_2}{R_1}(234.5 + T_1) - 234.5$$

or

$$T_2 = \frac{7.50}{6.24}(234.5 + 20) - 234.5$$

$$= 71.5^{\circ}$$

Hence the temperature rise above 20°C is

$$71.5 - 20 = 51.5^{\circ}C$$

Thus the average temperature rise of the coils is 51.5°C

From Table 6-1 we see that the temperature coefficients of resistance of most pure metals are nearly equal to 0.4 of one per cent change in resistance per degree change in temperature per ohm at 20°C. The temperature coefficients of alloys are much lower, some even zero or negative at certain temperatures. Manganin, an alloy of copper, nickel, iron and manganese has a resistivity in ohm-circular mils per foot varying from 250 to 450, depending on the proportions of the various metals, but its temperature coefficient is only 0.003 of one percent, practically negligible. Thus manganin finds its greatest use in instruments where fixed resistance is necessary. Certain substances, especially carbon, procelain and glass have large negative temperature coefficients of resistance, i.e., the resistance decreases as the temperature rises. Thus the cold resistance of a carbon lamp filament is almost twice the resistance of the heated lamp.

Since copper is the most commonly used conductor in the electrical industry the Bureau of Standards has prepared Tables, which have been adopted by the American Institute of Electrical Engineers, that give the resistance of 1000 feet of annealed copper wire of different standard sizes and at several temperatures. The standard sizes are designated by an A. W. G. (American wire guage) number as to diameter in mils, cross sectional area in circular mils, etc. The A. W. G. guage is also known as the B. & S. (Brown and Sharpe) wire guage. We may use these tables to find the resistance of any length of conductor of a given cross-sectional area. Table 6-2 lists data on some of the most commonly used sizes. It is possible to estimate the guage number, the resistance and size of any wire roughly if we keep in mind a few simple facts about the wire tables. No 10 wire is very nearly 0.10 inches (100 mils) in diameter, 10,000 circular mils in cross-sectional area and has a resistance very nearly one ohm per 1000 feet. As the wire diameter decreases, every third guage number doubles the resistance and halves the cross-sectional area. Thus No. 5 has about 33,000 circular mils and 3/10 of an ohm per 1000 feet while No. 8 has 16,000 circular mils and about 6/10 of an ohm resistance per 1000 feet. As the wires increase in diameter, every third guage number doubles the circular mils area and halves the resistance. No. 2 for example, has an area.of 66,000 circular mils and a resistance of 1.5/10 ohm per 1000 feet. The weight may also be roughly approximated from the fact that 1000 feet of No. 0 wire weighs very nearly 320 lbs. and the weight of a conductor varies inversely with the resistance. As an example we will choose the proper wire size to transmit electrical power for 4 miles if the resistance is to be less than 3.0 ohms at 20°C.

$$4 \text{ miles} = 4 \times 5280 = 11,120 \text{ feet}$$

$$3.0 \text{ ohms for 4 miles} = \frac{3.0}{11.12} \text{ ohm per 1000 feet}$$

$$= 0.272 \text{ ohm per 1000 feet}$$

From Table 6-2

$$\text{No. 4} = 0.248 \text{ ohm per 1000 feet}$$

$$\text{No. 5} = 0.313 \text{ ohm per 1000 feet}$$

Hence No. 4 wire will be used in order not to exceed 0.272 ohms per 1000 feet.

If the same line is to work at a temperature of 42°C what number wire will be required? Thus, first we must find the resistance at 42°C when we know the resistance at 20°C. The temperature rise above 20°C is 42°C - 20°C = 22°C. For each degree rise the resistance increases by 0.393 per cent. For a 22 degree rise the resistance has increased by 22 x 0.393 = 8.65 per cent. Hence at 42°C the resistance is 108.65 per cent of its value at 20°C.
The resistance at 42°C is

$$3 \times 1.0865 \doteq 3.26 \quad \text{ohms for 4 miles}$$

The resistance per 1000 feet $= \dfrac{3.26}{11.12}$

$$= 0.294$$

No. 4 wire may still be used since the limit of 0.294 ohm per
1000 feet will not be exceeded.

Stranded cables are very often used instead of solid wires.
Stranded cables have greater flexibility and are much easier
to pull into a conduit while less likely to break when bent
around sharp edges. Wire sizes greater than No. 0000 are generally specified by area such as 250,000 circular mils and
300,000 circular mils being standard and practically always
stranded for convenience in handling.

We have seen that an electric current will heat the conductors
through which it is flowing. The temperature will rise continuously in a conductor carrying current as long as heat is
generated faster than it can be dissipated from the surface of
the conductor. Therefore, in installing wires in buildings, the
wire sizes must be selected so that they will dissipate the heat
generated by the current at such a rate as to prevent the insulation from deteriorating.

Tables of Safe Current Capacity of the copper wire sizes used
in interior wiring are issued by the National Board of Fire
Underwriters. Thus the National Electric Code must be con-
sulted in designing wiring for commercial or residential buildings. The N. E. C. specifies the allowable current in conductors
of all sizes permitted in wiring installations, for the various
methods used in installation and for the different types of insulation available. Some data on allowable currents for the most
commonly used wire sizes is given in Table 6-2.

In choosing the proper wire size for any part of an interior
distributing system it is important to keep two facts in mind.
First, it is desired to choose the smallest wire size to carry
the required current without overheating to cause damage to
the insulation. Second, it is also desirable to choose the smallest wire size that will not cause excessive voltage drop. Therefore, the size in every section of the interior installation must
be such that no wire exceeds the current carrying capacity
specified by the N. E. C. Tables. Also, since most electrical
appliances will operate satisfactorily only at rated voltage, it
is necessary to compute the voltage drop throughout the entire

distribution system in order to maintain certain definite voltage levels. A variation of more than 5 per cent in the voltage
used for lighting produces an unpleasant fluctuation in illumination. For a total load including motors and heating appliances
a maximum voltage drop of 10 per cent is permissible, but any
larger drop would affect the speed of motors adversely.

A typical distribution diagram for a house wiring is shown
in Figure 37. The service company usually brings its wires to
a point called the service entrance where the main switch, protective devices and metering equipment is installed. From the
service entrance, heavy conductors are brought to the central
distribution point. Here the main panel board is installed. The
panel board provides a switch and protective devices for each
of the circuits required, known as feeders. The feeders are
usually brought to subpanel boards from which branch circuits
run to the various outlets.

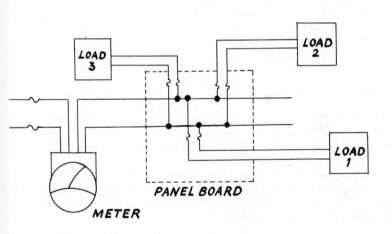

Figure 37

In Figure 37 the panel board is 100 feet from the main switch.
Three branch lines run from the panel board; each branch is
supplied with 10 outlets for 60 watt, 110 volt lamps. We shall
determine the wire size of the mains. Each branch line carries

10 x 60 or 600 watts. Each branch wire carries $\dfrac{600}{110}$ or 5.45 ampere. From Table 6-2 we see that No. 16 wire would meet this requirement adequately. Each main carries the current in all three branches; i.e., 3 x 5.45 or a total of 16.35 amperes. Hence No. 12 wire must be used.

It is now necessary to check the wire sizes to see if they meet the limitation of the voltage drop requirement, namely, less than 5 per cent required for proper illumination. For convenience in calculation, all lamps on each branch may be considered to be concentrated at a point on the branch line commonly known as the load center of the branch. In Figure 37 this distance is found to be 50 feet. Thus the length of wire in each branch is 50 x 2 or 100 feet.

The resistance of 1000 feet of No. 16 wire (Table 6-2) = 4.016		ohms
The resistance of 100 feet of No. 16 wire = 0.402		ohms
The voltage drop in each branch is 5.45 x 0.402 = 2.18		volts
The length of mains is 2 x 100 feet = 200		feet
The resistance of 100 feet of No. 12 wire = 1.59		ohms
The resistance of 200 feet of No. 12 wire = 0.318		ohms
The voltage drop in the mains is 16.35 x 0.318 = 5.19		volts
The total voltage drop from the mains to the lamps is 5.19 + 2.18 = 7.37		volts
The per cent line drop is $\dfrac{7.37}{110}$ = 6.7		per cent

Good practice indicates that this is more than the desirable percentage drop. Thus there would be a wide variation in brightness depending on the number of lamps in use. We should therefore try a No. 10 wire for the mains.

The resistance of 200 feet of No. 10 wire is 0.2 ohms

The voltage drop in the mains is 16.35 x 0.20 or 3.27 volts. The total voltage drop in the line is now 5.45 volts. Hence the percentage voltage drop in the line is

$$\frac{5.45}{110} = 4.95 \text{ per cent}$$

Thus the use of No. 10 wire for the mains very adequately meets the current carrying capacity and in addition provides a voltage drop in the line which is within the 5 per cent limit called for in good wiring practice.

Table 6-1

Resistivity and Resistance Temperature Coefficient

Material	Resistivity		Resistance temperature coefficient at 20° C
	microhm-centimeter at 20° C	ohm-circular mils per foot at 20° C	
Advance Metal (alloy)	48.8	298	0.000018
Aluminum	2.83		0.0039
Antimony	41.7		0.0036
Brass	7.0	40	0.0021
Carbon	3000.0		
Chromel	90.0		0.00008
Constantan	49.0		0.000008
Copper (Standard anneald)	1.724	10.37	0.00393
Copper (Hard drawn)	1.77	10.78	0.00382
Copper (Pure annealed)	1.692		0.0041
German Silver	33.0		0.0004
Graphite	800.0		
Iron, Commercial hard cast	98.0		
Lead	22.0		0.0039
Manganin	44.0		0.000006
Mercury	96.0		0.00089
Nichrome	99.6		0.00044
Silver	1.63		0.0038
Steel (Siemens-Martin)		98.1	
Steel (4% silicon)	50.0		0.0008
Steel (rails)	13.8-21.6		
Tungsten	5.15	33.2	0.0045

Table 6-2 Annealed Copper-Wire Data

AWG Number	Diameter in mils	Area in circular mils	Resistance in ohms per 1000 feet	Allowable current in amperes (Rubber insulated in cable or raceway)
0000	460	212,000	0.050	195
000	409.6	168,000	0.063	165
00	364.8	133,000	0.078	145
0	324.9	106,000	0.100	125
1	289.3	83,700	0.124	110
2	257.6	66,400	0.156	95
3	229.4	52,600	0.197	80
4	204.3	41,700	0.249	70
5	181.9	33,100	0.314	55
6	162.0	26,250	0.400	50
8	128.5	16,500	0.630	40
10	101.9	10,400	1.000	30
12	80.8	6,530	1.590	20
14	64.0	4,110	2.525	15
16	50.8	2,580	4.016	6
18	40.3	1,620	6.385	3
20	32.0	1,020	10.150	
24	20.1	400	25.67	
28	12.6	160	64.9	
32	7.95	63.2	164	
36	5.0	25	415	
40	3.15	9.9	1,050	

CHAPTER 7—MAGNETIC CIRCUIT CONCEPTS

7.1 The Magnetic Field

Almost all electrical equipment, particularly electric power machinery is dependent upon the interaction of electricity and magnetism for its operation. As early as 1820 Hans Christian Oersted observed that a magnetic compass which was placed near a wire carrying an electric current experienced a deflection which was definitely due to the current. Ampere checked this observation and found that mechanical forces exist between different conductors carrying different current as well as between different portions of a conductor carrying the same current. In 1831 Michael Faraday discovered the principle of electromagnetic induction which led to the invention of the electromagnetic generator upon which our present day electrical technology depends. We may summarize the findings of these men in terms of three magnetic effects which are due to a flow of current:

1. Forces are experienced by magnets and ferromagnetic materials in the neighborhood of the current.
2. Forces are experienced by nearby conductors in which currents are flowing.
3. Currents are induced to flow in other conductors forming closed circuits whenever there is a relative motion between the circuits, or the strength of the current which is the cause of the effect is varied.

Thus whenever an electric current flows in a conductor in addition to producing heating and possible chemical effects it makes its presence felt even in the space about the conductor. These effects, known as magnetic effects may be observed both inside and outside of the conductor. We therefore assign to any region of space, in which a wire experiences a mechanical force due to the current which it is carrying, or in which an electromotive force is induced in a circuit due to its motion in the region, the property of a magnetic field. Any region in which a magnet is subject to mechanical forces is also known as a magnetic field. The magnetic fields that we are concerned with in the study of electrical technology are usually produced by a conductor which is carrying a current or a permanent magnet.

Our modern theory of magnetism is based upon currents

which circulate within the atoms and hence upon current flow so
that we may conclude that all magnetic phenomena are due to the
movement of electrical charges, i.e., current flow. The concept
of a magnetic field involves the idea of magnetic flux, which really
represents the condition of any region of space, not any physical
entity. However magnetic flux, may be conveniently represented
by lines, just as stream lines are used to describe the flow of
water. Thus a convenient means of describing the magnetic
effect is the use of lines to describe the direction and intensity
of magnetic flux.

The familiar iron-filing experiment lends a certain amount
of credence to the concept of the magnetic field and its accom-
panying magnetic flux. A long wire is passed through a hole in a
card. The card is then sprinkled with iron filings. A current
is made to flow in the wire and then the card is tapped gently.
The iron filings arrange themselves in circles. This proves that
the magnetic field is circular about the wire since it is known
from previous experiments that all ferromagnetic substances,
when placed in a magnetic field, tend to arrange themselves in
such a position that their longest dimension is parallel with the
direction of the magnetic field. The direction of the magnetic
flux is determined by the direction of the current in the wire.
The following procedure has been found useful in finding the
direction of the magnetic flux. If the conductor is grasped in the
right hand and the thumb is extended in the direction of the current flow in the conductors, the fingers will point in the direction of
the flux.

7.2 The Magnetic Circuit

In electrical power equipment such as generators, motors and
transformers, the magnetic field is usually set up in an iron core
and in the narrow air gaps in this core. The core, in general, has
a substantially uniform cross-sectional area, throughout most of
its length, in which magnetic flux is largely confined. There will
be, unavoidably, a magnetic field outside of the core and the
air gaps, a non useful field, since it is not essential for the operation of the electrical equipment. Such a field is called a leakage
field and the flux associated with this field is called leakage flux.
Electrical conductors in the form of coils are wound over this
core. The core and its air gaps are considered a magnetic circuit
An electromotive force can be arranged to produce an electric
current in a metallic circuit. Similarly, a coil, wound over an iron
core, carrying an electric current will cause magnetic flux to be set u
in the iron core. Since the current in the coil is the cause of the mag-

netic flux, we may refer to it as a source of magnetomotive
force responsible for causing the flux to be set up in the core.
The magnetomotive force, the core and the flux set up in it may
be called a magnetic circuit, similar to the electrical circuit.

We shall see that there is a great similarity between the two
circuits and we shall find that the laws governing the magnetic
circuit may be stated in much the same form as those for the
electric circuit. There are very significant differences, however,
which must be emphasized. A conductor which carries current
is normally isolated from other conductors by insulating materials. Rubber insulation, for example, has a conductivity of
nearly 10^{-20} times that of copper, resulting in a negligible current flow in the rubber insulation. No such magnetic insulator
is available. In addition an air gap in the core, often desired in
the magnetic path, is in parallel magnetically with another air
path having at most five or ten times the magnetic insulating effect of the air gap. Thus a considerable percentage of the magnetic flux is shunted from the useful path into the parallel air
path where it does not serve any useful purpose. This is the condition which we have defined as leakage flux.

In an electric circuit, the flow of current results in heat
dissipation in the conductor even when direct current flows
without any change in magnitude. A magnetic circuit, in con-
trast, will produce no heating of the magnetic circuit because of
a steady magnetic flux, no matter how long the flux exists in the
circuit. This of course means that in so far as the magnetic
circuit is concerned, no energy is being dissipated. Finally, we
stress once again the fact while in an electric circuit the current
is defined as the movement of charge across any cross-section
of the circuit, we do not believe that there exists any net physical
motion of anything across any cross-section of the magnetic circuit.

7.3 Measurement of Magnetic Flux

Faraday's original experiments in electromagnetism are of
principal importance in yielding information about the manner
in which magnetic flux varies when it is set up in a magnetic
circuit. In one of his earliest experiments he designed a magnetic circuit by winding two coils of insulated wire upon an iron
ring (core). He connected one of the coils to a battery in series
with a switch which enabled him to open and close the circuit at
will. The second coil was connected to a galvanometer, an instrument which is sensitive to current flow. A simplified picture of
Faraday's magnetic circuit is shown in Figure 38. Upon closing the

switch and thereby energizing circuit one he observed a momentary deflection of the galvanometer in circuit two. Thus a current
was seen to flow momentarily in the second coil. As soon as the
switch was opened, thereby, breaking the circuit in coil one a

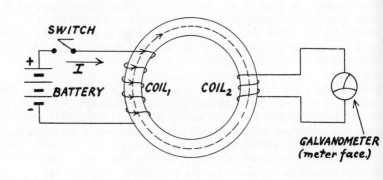

Figure 38

deflection of the galvanometer was observed but in the opposite
direction. Since there is no electrical connection between the
coils, Faraday reasonably assumed the magnetic flux which was
set up by the current in coil one in the iron core, had something
to do with producing the current, as evidenced by the deflection
of the galvanometer, in coil two. However, the experiment
showed that as long as the current in coil one remained constant
there was no deflection of the galvanometer indicating that there
was no current flow in coil two. He therefore concluded that as
long as the current flow was constant the magnetic flux also remained constant and no effect is observed. But a change in the
current in coil one must be accompanied by a change in the number of magnetic flux lines and that this change is responsible for
setting up an electromotive force in coil two. It thus appears
that the magnetic flux produces an effect, i.e., introduces an
emf in circuit two, while it is changing, an emf in one direction

while it is being set up and an emf in the opposite direction
while it is decreasing. We can study these effects experimentally by replacing the switch, in coil one, by an ammeter and a
rheostate in series with the coil and the battery. Thus we may control the current in coil one by keeping it at a steady value or by
increasing it or decreasing it, by changing the total resistance of
the circuit when we change the position of the rheostate. We find
that as long as the ammeter shows us that a steady current is
flowing in coil one there is no emf induced in circuit two. When
the ammeter indicates that the current is increasing, the gavan-
ometer deflects in one direction; and when the ammeter indicates that the current is decreasing, the galvanometer deflects
in the opposite direction. We make the current change quickly
and slowly by adjusting the rheostat at different rates, and observe that the emf which is induced in the second coil is exactly
proportional to the rate at which the flux is varying in the magnetic circuit, i.e., the rate at which the current is changing in
circuit one. It seems logical, therefore, to evaluate the induced
emf as a measure of the flux which is produced in the magnetic
circuit by the changing current in coil one. To measure the
rate at which the flux is changing in the magnetic circuit, the
time that it takes to make the change and the emf which is induced in the second coil is difficult and not practical. In practice, the measurement of magnetic flux is achieved by the use
of a flux-meter or a ballistic galvanometer. A ballistic galvano-
ometer differs from an ordinary current indicator in that due to
its very sluggish action it will give the maximum swing which
occurs due to a momentary current. Thus, a ballistic galvanometer may be used to indicate the amount of a voltage impulse
which is suddenly applied to its terminals. That is, when a potential is applied to a ballistic galvanometer for a very short
time and then removed, the deflection of the meter will be found
to be proportional to the product of the voltage applied and the
time that it was acting on the circuit. Since a voltage impulse
represents a product of voltage and time, it is measured in volt-
seconds. If a constant potential E is applied for a short period
of time ΔT the voltage-time product is found to be directly
proportional to the deflection and may be stated mathematically
by the equation

$$D = K_1 \ E \ \ \Delta T$$

where D is the galvanometer deflection, K_1 is a constant of the
galvanometer and test circuit. Faraday's experiment for electromagnetic induction may be generalized in the following way:
The emf induced by a changing magnetic flux in a magnetic cir-

cuit is proportional to the rate at which the flux changes; or
stated mathematically

$$E = K_2 \frac{\Delta \phi}{\Delta T}$$

where E is the average emf in volts,
$\Delta \phi$ is the change in flux,
ΔT is the time for the change of flux to occur, and K_2 is the constant of proportionality. Therefore the deflection of the ballistic
galvanometer is

$$D = K_1 K_2 \Delta \phi$$

or $$D = K (\phi_A - \phi_B) \quad \text{where } K = K_1 K_2$$

This shows us that the deflection of a ballistic galvanometer is a
measure of the amount of change in flux in a magnetic circuit, ϕ_A
representing the initial value of the magnetic flux and ϕ_B the final
value. Hence for a practical measurement of magnetic flux we
may set up the following circuit, similar to the one shown in
Fig. 39. We place an ammeter and reversing switch in the first
coil, which we now call a magnetizing coil, since it produces the
flux in the magnetic circuit, the second coil of only a few turns
is terminated in a ballistic galvanometer and now is referred to
as the measuring coil. The measurement may now be made as
follows:
   When the current is suddenly changed in the magnetizing coil
the resultant change in the amount of flux in the magnetic circuit
may be found from the deflection of the galvanometer which is
proportional to the total amount by which the flux is changed. For
an iron core magnetic circuit, the initial deflection when the
flux in the circuit is first set up, will not be a measure of the
flux in the magnetic circuit unless we are sure that there was
no flux in the iron core before we energized the magnetizing
coil. This is due to a property of the iron which we shall discuss
fully later. By means of the reversing switch, which we move
back and forth several times, we assure that the initial deflection
is a measure of the amount of flux in the circuit. By reversing the
switch in the magnetizing circuit we are reversing the current
from a maximum in one direction to a maximum in the other
direction hence the flux in the magnetic circuit is changed from a
maximum in one direction to a maximum in the other. Therefore,
by dividing the galvanometer deflection by two we obtain the maximum amount of the flux in the magnetic circuit.

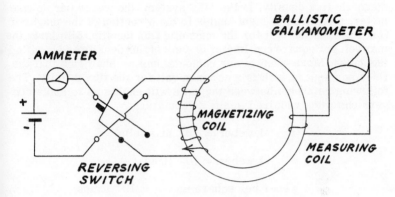

Figure 39

We define the unit of magnetic flux, in the MKS system, as the weber. When in a magnetic circuit, an amount of flux is set up such as to produce in a coil of one turn, wound on the magnetic circuit, one volt-second we define this amount of flux as the weber. We may also say that a voltage-impulse of one volt-second is induced in a coil of one turn when the flux through the coil is changed by one weber. The weber, or unit of flux, is really defined in terms of a change of flux. Only a changing flux is capable of inducing a voltage in a coil which links the magnetic circuit. Although the MKS system of units is now the standardized system, a mixed set of English units are found in practice, which prove convenient when dimensions of electrical equipment are given in inches. Thus, the maxwell, the maxwell per square inch and the ampere-turn per inch are commonly used. We shall simply relate these units to the MKS magnetic units to facilitate computation. The maxwell or line is a unit of flux and is related to the weber by the following equation

$$1 \text{ weber} = 10^8 \text{ maxwells}$$

$$= 10^8 \text{ lines}.$$

When we deal with magnetic flux, the density of the flux lines is sometimes more important than the flux itself. In our equations and computation of magnetic circuits we will often use the magnetic flux density. In the MKS system, the weber per square meter, measured at right angles to the direction of the magnetic flux lines, is the unit for the magnetic flux density. Similarly the maxwell per square centimeter is a unit of flux density and is called the gauss. More often in our computations we shall find that the lines per square inch is a common unit for the flux density. The following relations between the units will enable us to make conversions when required in any computations.

$$1 \text{ weber} = 10^8 \text{ maxwells}$$

$$1 \text{ weber} = 10^8 \text{ lines}$$

$$1 \text{ weber per square meter} = 10^4 \text{ gauss}$$

$$1 \text{ gauss} = 1 \text{ maxwell per square centimeter}$$

$$1 \text{ maxwell per square inch} = 0.155 \text{ gauss}$$

$$1 \text{ kiloline} = 10^{-5} \text{ webers}$$

1 kiloline per square inch = 0.0155 webers per square meter

Consider that the flux through a coil of one turn is being increased at a uniform rate from zero to 120,000 lines in 3 seconds. What will be the average emf induced in volts?

$$\Delta\phi = 120,000 \text{ lines} = 120,000 \times 10^{-8} \text{ webers}$$

$$\Delta\phi = 12 \times 10^{-4} \text{ webers}$$

where $\Delta\phi$ represents the change of flux. Then, E, the average emf induced in the coil will be

$$E = N \, \frac{\Delta\phi}{\Delta T} \quad \text{volts}$$

Here N is the number of turns on the coil and $\frac{\Delta\phi}{\Delta T}$ represents the average rate of change of flux.

Thus
$$E = \frac{12 \times 10^{-4}}{3} = 0.0004 \text{ volts}$$
$$= 400 \text{ microvolts.}$$

If the coil has 500 turns, what will be the average emf in the coil. Since an emf is induced in any coil with which the changing flux is interlinked, 500 turns of wire may be considered as a single turn linking the changing flux 500 times hence the average voltage induced in a coil of 500 turns will be 500 times that induced in a single turn. Hence the average emf is

$$E = 500 \times 0.0004$$
$$= 0.2 \text{ volts}$$

Thus we may summarize the phenomenon of electromagnetic induction by stating that when flux is changing through a coil an emf is induced in the coil. We have defined the weber as that amount of flux which will produce a one volt-second impulse in a single turn coil. Hence, if the flux through single turn coil is changing at the rate of one weber per second, one volt will be induced in the coil. Therefore we conclude that the emf induced in a coil is equal to the rate of change of the flux linkages through the coil. The mathematical equation we used above enabled us to compute the average emf induced in a coil due to an average rate of change of flux, i.e., $\frac{\Delta \phi}{\Delta T}$.

From the calculus we know that the

$$\lim_{\Delta T \to 0} \frac{\Delta \phi}{\Delta T} = \frac{d\phi}{dt}$$

Therefore, the instantaneous voltage induced in the coil will be

$$e = - \frac{N \, d\phi}{dt} \text{ volts}$$

Of course, the instantaneous value and the average value will be identical if the rate of change of flux is uniform. The minus sign appears on the right hand side of the equation and is the result of an experimentally determined relation by H. Lenz. The experimental researches of Lenz led him to formulate the electromagnetic relation as follows:

The electromotive force induced in an electric circuit linking a changing magnetic flux, is directly proportional to the time rate of change of the flux linkages and tends to produce a current in such a direction as to oppose the change in the flux. This relationship has been termed Lenz's law.

7.4 Magnetomotive Force

We have already seen that a coil of wire carrying a current when interlinked with a magnetic circuit may be considered as a magnetomotive force tending to set up a magnetic flux in a magnetic circuit. The name magnetomotive force is a result of the analogy of the magnetic circuit to the electric circuit where a battery may provide the electromotive force which causes current to flow around the circuit. There is however, a very significant difference between the electromotive force and the magnetomotive force. On the one hand the battery is an integral part of the electrical circuit, producing the emf directly in the circuit by chemical action. The mmf, on the other hand, in a magnetic circuit, is furnished by a coil of wire carrying a current which does not physically touch the magnetic circuit, that is, there is no direct connection. The only requirement is that coil surround or link some part of the magnetic circuit.

If a magnetizing coil is wound on a core which does not contain iron or any magnetic material and a current is made to flow in the coil, flux will be set up in the core, which is referred to as a non-ferrous magnetic circuit. Experimentally it is observed that the total magnetic flux is proportional to the current in the magnetizing coil and also to the number of turns of wire used in the coil, i.e., the number of times the current encircles the flux path. If all the turns are wound in the same direction about the magnetic circuit we may conclude the magnetomotive force, mmf, is proportional to the product of the current flowing in the coil by the number of turns of the coil. In general, the mmf, which is given the symbol F, is proportional to the product of the current and the number of linkages between the magnetizing coil and the magnetic circuit.

In measuring the mmf, as in the measurement of magnetic flux, various system of units have been in use. The MKS system expresses the unit of mmf directly in ampere-turns. Thus, the magnetomotive force of a coil in ampere-turns, is the product of the current in amperes by the number of turns in the coil. Expressed mathematically,

$$F = NI \quad \text{ampere turns.}$$

We have already stressed the analogy of the mmf of a coil to the
emf of an electric circuit. The emf was considered as a rise
in the electrical potential through the source, while current flowing through a wire is always associated with an electrical potential difference between the ends of the wire. We may now extend
the analogy and say that a magnetic potential difference exists
over the length of the flux path whenever there is magnetic flux
in an iron or non ferrous material. If the ends of the flux path
are bent so that they touch, the mmf, which is necessary to produce the flux, would be exactly balanced by the magnetic potential
drop over the length of the flux path. Thus the magnetic potential
difference is equal to the mmf which is necessary to maintain
the flux in a particular part of the magnetic circuit. For any magnetic circuit the sum of the total magnetic potential drops
around any closed path in the magnetic circuit is equal to the
mmf which is acting on that closed path. When the mmf is expressed in ampere turns, the sum of the total magnetic potential
drops around any closed path is equal to the ampere turns en-
closed by that path. This result is the equivalent of the Kirchhoff
voltage law for electric circuits.

The design of electrical power equipment usually involves the
problem of determining the currents necessary to produce a
desired magnetic flux across a surface. In such computations it
is frequently convenient to use a quantity which represents the
magnetic potential drop per unit length. This is especially true
when working with ferrous magnetic circuits where in general
the mmf's will not be uniformly distributed over the entire length
of the magnetic circuit. The MKS unit for this quantity, which is
given the symbol, H, is the ampere-turn per meter. The ampere-
turn per meter, H, is also known by two other names, namely, the
magnetic field intensity and the magnetizing force. It is often expressed in ampere-turns per inch or in oersted. For convenience
in computation the relation between the units commonly found in
practice is listed.

1 ampere-turn = 1.257 gilbert

1 gilbert   = 0.796 ampere-turn

1 oersted   = 1 gilbert per centimeter

1 ampere turn per inch = 0.495 oersted

1 ampere-turn per meter = 0.001 oersted

                        = 39.37 ampere turns per inch.

7.5 The Magnetic Properties of Ferrous Materials

The use of ferromagnetic materials, iron and its alloys, is essential in producing strong magnetic fields having the definite geometric patterns usually required for many electromagnetic devices. It is important, therefore, to become familiar with the magnetic properties of these substances. The most useful physical characteristics of iron and the other ferromagnetic materials are the magnetization curves and hysteresis loops. A description and explanation of these magnetic properties are important in order to guarantee their optimum utilization in electrical technology.

7.5a Magnetization Curves

For the non-ferrous materials a simple straight line relationship exists between the magnetic flux through a piece of the material and the magnetic potential drop across it at any given temperature. No such simple relation is found for the ferromagnetic substances. Thus when a plot is made of the flux through a piece of iron as a function of the magnetic potential drop across it, the relation is a non linear one and is affected by the previous magnetic, thermal and mechanical history of the sample used. A typical curve is shown in Fig. 40. The curve applies to only one piece of a certain type of iron. If we desire to have a plot of this relationship which will apply to any sample that may be made of this type of iron we plot, the flux density, the flux per unit cross-sectional area as a function of the magnetizing force, the ampere turns per unit length. A plot of this type is known as a B-H curve.

Thus the relation of the magnetic flux density, which has been assigned the symbol B, to the magnetic field intensity, H, is non linear for ferromagnetic materials and is usually given in graphic form. Figure 41 represents a set of curves, plotted to give this relation between B and H for some of the different types of iron commonly used in magnetic circuits. These curves are averages of many samples of the particular type of iron and do not necessarily apply to any particular sample. We give them here in order to get an appreciation for the relative magnitudes of these quantities found in actual use. They may of course be used in the solution of the magnetic circuit problems at the end of this book. The designer of a piece of electrical machinery has at his disposal the actual B-H curve for the particular sample of the material which will be used in the machine. The computations will, of course, be based on these actual curves and not on any

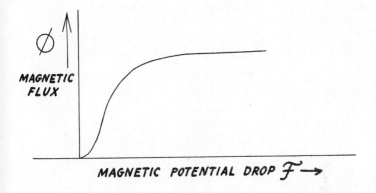

Figure 40

average curves. As is shown in Figure 41, for practical application these curves are plotted in kilolines per square inch for
the flux density, B, and in ampere-turns per inch for the magnetizing force. The B-H curve is known as the magnetization
curve. The magnetization curve finds its greatest use in determining the number of ampere-turns that are required to produce
a definite amount of flux in any particular magnetic circuit. If
we start with a sample of iron which has been magnetized, in
plotting the relation between B and H we observe that at first as
H is increased from zero, B is also increased and is seen to increase rather rapidly with H up to the "knee" of the curve. The
"knee" can be seen to be that portion of the greatest curvature
lying just above the steepest part of the curve. Thereafter B increases very slowly with H approaching a constant value as
the curve is extended. The part of the magnetization curve above
the beginning of the "knee" is commonly referred to as the region
of saturation and when the region of the upper straight part is
reached the material is said to be magnetically saturated.

7.5b Permeability and Reluctance

For non-ferrous materials, including air, the B-H characteristic was found to be linear, the slope of this straight line is a

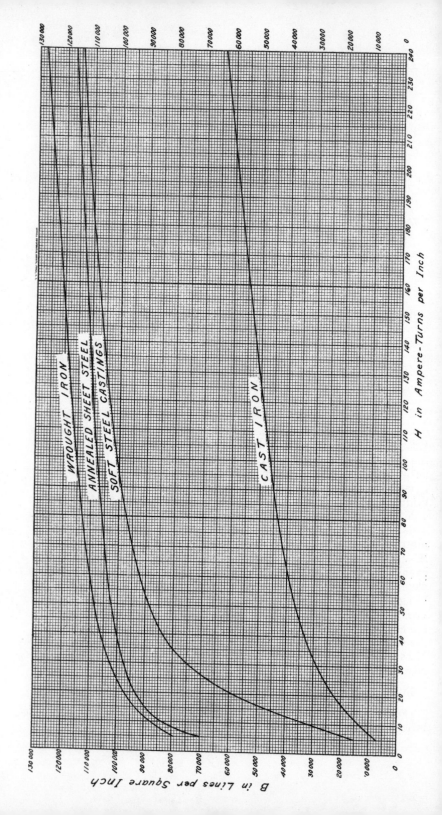

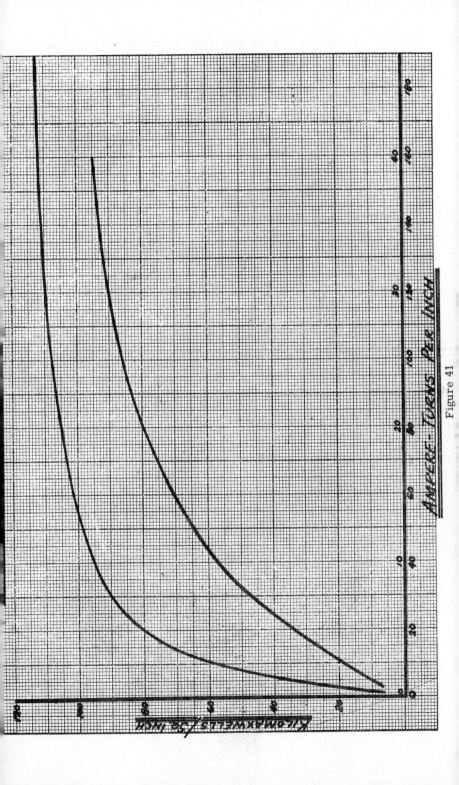

Figure 41

constant whose value is found to be $4 \pi \times 10^{-7}$ webers per
meter-ampere-turn, when B is in webers per square meter and
H is measured in amper-turns per meter. This constant represents the relationship of the magnetic flux density to the magnetizing force for free space and has been defined as the permeability of free space. Thus for free space and non-ferrous substances.

$$\frac{B}{H} = \mu_0 = 4 \pi \times 10^{-7} \text{ webers per meter-ampere-turn.}$$

The magnetization curve for iron also enables us to get a
relation between B and H. The ratio of B to H is called the
permeability of the iron and is represented by the symbol,
(the Greek letter mu). Thus the permeability of a material is,

$$\mu = \frac{B}{H} \quad \text{and represents the slope of the}$$

B-H curve and is different at each point of the curve for the
ferrous materials used in magnetic circuits. All the ferromagnetic materials have permeabilities very much greater than μ_0.
The ratio of the permeability of any material to that of free
space is a convenient value for specifying the relative permeability of the ferrous materials. The ratio is generally symbolized by μ_r and is seen to be a dimensionless value; thus

$$\mu_r = \frac{\mu}{\mu_0} = \frac{B}{\mu_0 H}$$

One of the materials shown in Figure 41, cast iron is quite in-
ferior as a magnetic material. However, it is quite extensively
used where space or strength is not a deciding factor because it
is inexpensive and easy to cast and machine. For low values of
H, the B-H curve for cast iron is practically a straight line. The
relative permeability is very nearly constant at about 350. At
higher values of H, the gain in flux density is less pronounced
and relative permeability decreases gradually. Soft steel casting
is superior to cast iron magnetically and structurally and is frequently used where intricate shapes are to be cast. At low flux
densities its relative permeability is very nearly 550, as H increases, the relative permeability increases to about 1500 and
as the steel saturates, decreases.

Silicon sheet steel finds extensive use in a-c and d-c motors
and generators and in transformers, where the exciting current

is continuously changing. Its initial relative permeability is 1200 and its maximum is nearly 6000. Certain nickel-iron alloys are capable of developing extremely high flux densities at relatively low magnetic intensities, reaching maximum relative permeabilities as high as 100,000.

The ratio of the magnetic potential drop to the flux in any part of a magnetic circuit yields an additional quantity, especially interesting, because of its analogy to the electric circuit. We define this ratio as the reluctance of that part of the magnetic circuit. Thus, the reluctance, symbolized by R is

$$R = \frac{F}{\phi} \quad \text{ampere-turns per weber.}$$

If the part of the magnetic circuit for which we have formed the ratio has a definite cross-section, A square meters and a definite length ℓ meters, and if the flux density, B, may be considered constant over this cross-section A while, H, the magnetizing force may considered constant along the length ℓ then:

The flux will be given by

$$\phi = BA$$

and the mmf will be

$$F = H\ell$$

whence the reluctance will become

$$R = \frac{H\ell}{BA}$$

or

$$R = \frac{\ell}{\mu A}$$

The reluctance of a magnetic circuit can thus be seen as analogous to the resistance of an electric circuit, since the resistance of a conductor may be defined in terms of its demensions

as, $R = \frac{\ell}{\gamma A}$ where γ is the conductivity of the conductor. The

reluctance, R, which we have defined as the ampere-turn per weber, i.e., a magnetic flux path will have a reluctance of one

unit in the MKS system if a magnetomotive force of one ampere-turn produces a flux of one weber, is seen to be a function of the permeability of the magnetic circuit. The defining
equation:

$$R = \frac{F}{\phi}$$

is often referred to as the "Ohms" law for magnetic circuits
from the similar expression for the electric circuit

$$R = \frac{E}{I}$$

The analogy is good for air and other non-ferrous materials
where μ_O, the permeability is a constant. However, magnetic
circuits are made from iron or other ferromagnetic material,
whose permeability is a function of the flux density and, therefore, the reluctance, which is a function of the permeability is
not a constant. In the electric circuit, as long as the temperature is reasonably constant, the resistance is also constant.
Therefore, the "Ohms" law for the magnetic circuit is not as useful as we found the Ohm's law for the electric circuits. Hence
reluctance is generally useful in a qualitative way.

7.5c Hysteresis

We have seen the relation between the magnetic flux density
and the magnetic intensity, i.e., B and H, is a non linear one for
ferromagnetic material. When a graph of the variation of the flux
density with corresponding values for the magnetizing force
throughout a complete cycle of positive and negative values of the
magnetizing force is plotted it is found that not only is the relation
non linear, it is also multi-valued. To study the resultant curve
for cyclic magnetization we choose, as our sample, a torroidal
ring made of iron and uniformly wound with a coil, as is shown
Figure 42.

This magnetic circuit has the mmf uniformly distributed
around the ring thereby confining the flux practically all to the
ring and resulting in very little leakage. Thus the B-H characteristic obtained will be for the iron only. First, we make sure
that the iron is completely demagnetized, so that when there is
no current in the coil, there is no magnetic flux in the ring. We
now start the current at zero and increase it gradually, thereby
increasing H, which is directly proportional to the current; at the

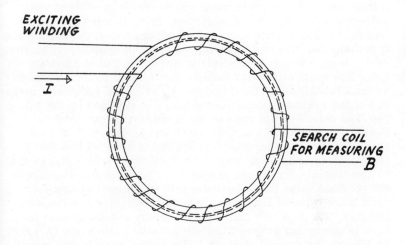

Figure 42

same time by means of the measuring circuit and the fluxmeter
readings we compute the corresponding increase in the flux
density, B. When a maximum current, arbitrarily selected is
reached, the B-H curve obtained is that from 0 to a in Figure
43a. Now we begin to reduce the current and H decreases, the
flux density B, decreases but not as rapidly as H. We observe
that when H is reduced to zero, B still remains very nearly
three quarters of its maximum value. Now we reverse the direction of the current, thereby making H negative, B begins to drop
rapidly until when H reaches the same maximum value as before,
but negative of course, B has also reversed and the B-H characteristic is now represented by the curve a-b. A reversal of H
and increase to + H_{max} once again brings us along the curve
b-c. Continued successive reversals from + H_{max} to - H_{max}
to + H_{max} results in a path which begins to be retraced as in
Figure 43.b, when a cyclic state is reached in the B-H characteristic.
Thus we observe that the iron exhibits a tendency to retain the
flux density which is set up in it by a magnetizing current. This
characteristic is known as magnetic hysteresis. Magnetic
hysteresis describes that property of a ferromagnetic material
whereby the substance, when completely demagnetized, and ex-

posed to a magnetizing force which is then removed, retains a certain amount of residual magnetism. Thus the iron sample, did not return to its previous magnetic condition after it was subjected to the cyclic magnitization. From this we conclude that the magnetic state of iron does not depend on the magnitizing force only but also on its previous history of magnitization. The cyclic loop, which the iron finally exhibits is called a hysteresis loop. The value of the flux density, B_r, when H is zero is known as the residual magnetic flux density and can be taken as a measure of the residual magnetization which is retained by the sample. The retentivity of the sample is the largest value of the residual magnetic flux density which can be obtained. If we desire to reduce the magnetic flux density to zero we find that it is necessary to apply a reversed magnitizing force - H_c, known as the coercive force for the particular value of B_r. The largest possible value of the coercive force, obtained by using a very large value of H_{max}, describes the coercivity of the iron. Residual flux densities and coercive forces are indigenous to ferromagnetic materials. Whenever a coil or a conductor carries an alternating current in electrical machinery cyclic magnitization takes place, as in the core of a transformer in service or in the iron near the windings of any alternating current generator or motor. In direct current motors and generators, the armature rotates in the magnetic field set up by the stationary poles, resulting in cyclic magnetization in the magnetic circuits. As the armature rotates a particle of iron in any part of it is being magnetized first in one direction, then in the opposite.

Figure 43c is a graph of B versus H for an alternating magnetizing force applied to a sample of iron. Here the maximum value of the magnitizing force has been adjusted to several different values in order to get the series of hysteresis loops corresponding to the different values of the maximum flux density of each loop. The positive peaks of these hysteresis loops may be joined and the resultant curve represents the normal or mean magnetization curve of the sample of iron used.

The suitability of a given ferromagnetic sample for a particular application can be determined from the size and shape of the hysteresis loop. Standard practice has established the determination of hysteresis loops by using a maximum flux density of one weber per square meter (64.5 kilolines per square inch). Therefore when B_r or H_c are given for a particular material, except permanent magnets of very high coercivity, it is assumed that the maximum flux density was one weber per square meter. In electrical power machinery we are usually concerned with periodically varying magnetizing forces and

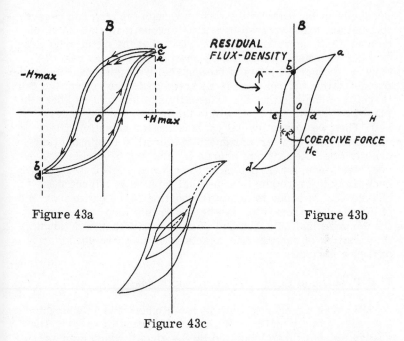

Figure 43a   Figure 43b

Figure 43c

therefore the mean magnetization curve will be of prime interest in determining the relation between the flux density and the magnetizing force. Flux densities above 2 webers per square meter (129 kilolines per square inch) are considered high flux densities. Experimentally, it is found that when flux densities become high, the slope of the mean magnetization curve approaches a value equal to μ_o, the permeability of free space.

Ferromagnetic phenomena, such as cyclic magnetization, hysteresis, retentivity can be explained in terms of circulating currents within each atom of the ferromagnetic material. In terms of our modern electron theory, these circulating currents are the result of electron motion, in a predominating pattern, in the molecules including such phenomena as the spin of the electron as well as its orbital motion. When an external mmf is applied to a ferromagnetic material, these circulating currents are said to orient their axes as nearly parallel to the axis of the magnetizing coil as possible. The currents will then be aiding the main magnetizing coil. As the magnetizing current increases, the axes of the atomic currents become more nearly parallel with the axis-of the coil resulting finally in a perfect allignment. When this occurs any additional increase in magnetizing current produces an increase in the flux as though the iron had ceased to exist and was replaced by a non-ferrous material. This, of course, finds corroboration in the fact that the mean magnetization curve at high flux densities shows a slope very nearly equal to the permeability of free space. It is this condition in the magnetization of iron that is referred to as saturation. We can therefore conclude that the externally applied mmf has been replaced by an internal mmf within the ferromagnetic material itself. Thus we have a logical explanation for retentivity, the characteristic of a ferromagnetic material to retain the flux density when the external source of mmf has been removed. A coercive force, a negative external magnetic force is therefore necessary to overcome this internal mmf as is indicated in Figure 43b, in order to obtain the portion of the curve from b to c. The section from b to c of the hysteresis loop is known as the demagnitization curve and is most significant in determing the behavior of permanent magnets when they are used as a part of a magnetic circuit.

7.5d Hysteresis Loss in Ferromagnetic Materials

Whenever a magnetic field is established it is accompanied by a storage and expenditure of a definite amount of energy. The energy, is, of course, supplied by the circuit which acts as the source of the mmf. For a magnetic field established in free space, all of the energy which is stored is returned to the circuit when the field collapses, as the current stops flowing. Generally, however, some of the energy is lost in space or the material in which the field is set up and hence is not recoverable. The amount of energy which is stored in a magnetic field can be expressed in terms of the magnetizing force, the magnetic flux density and the volume of space which is occupied by the field. The energy stored in the field can also be found to be given by the shaded area under the B-H curve as shown in Figure 43d. For free space, the B-H curve will be linear and the area will given by $1/2\ B_{max}\ H_{max}$. The energy thus found represents the energy stored in the magnetic field for each unit volume of space occupied by the field. The energy stored per unit volume in the iron will be given by the shaded area under the B-H curve and will be expressed in joules per cubic meters, when B is in webers per square meter and H is in ampere-turns per meter. The area under the curve may generally be found approximately by graphical methods. The calculus provides us with a ready means of expressing this area. Thus, the energy per unit volume of the iron is given by:

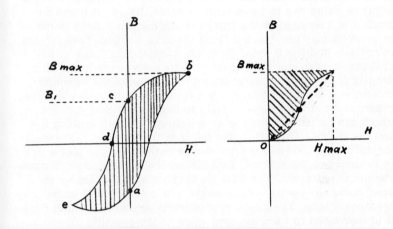

Figure 43e   Figure 43d

$$W = \int_0^{B_{max}} H\,dB \quad \text{joules/cubic meter}$$

For free space

$$W = \frac{1}{2}\,B\,H$$

and since $B = \mu_0\,H$

$$W = \frac{1}{2}\frac{B^2}{\mu_0} = \frac{1}{2}\mu_0\,H^2 \quad \text{joules/cubic meter}$$

When the flux density is carried through a hysteresis loop in a
ferromagnetic material, energy in the form of heat is dissipated
in the iron. Experimentally, it is readily observed that when a
ring of iron is exposed to an alternating magnitizing force, at 60
cycles per second or higher, the temperature of the iron rises
above the ambient. The expenditure of energy is generally
accounted for by the magnetic orientation of the axes of the iron

crystals parallel to the magnetizing force and therefore this energy loss is said to be due to the hysteresis and referred to as hysteresis loss. In Figure 43e while H is increased from zero to its maximum value, the flux density will be changing from a negative value o-a to its maximum value, B_{max}, at b and an amount of energy is being stored in the magnetic field in the iron which may be obtained from the area included between the curve a-b and the B axis. Now, the magnetizing force is decreased to zero and is accompanied by a decrease in the flux density from its maximum value to the value of the residual magnitism, B_r, from b to c along the curve and energy is being returned from the magnetic field to the magnetizing circuit, which can be found from the area between the curve along b-c and the B axis. The difference between these areas represents the difference between energy stored in the iron and that which is returned to the source. This difference is, of course, the net amount of work which is done on the iron and is given by the area under the portion of the curve a-b-c or one half of the hysteresis loop. The above procedure is repeated but this time H is increased to its maximum value in the opposite direction, and then decreased to zero again. We may now compute the net amount of work which is done on the iron when the flux density is made to vary from its maximum value in one direction to its maximum in the other, since the reversal of the direction of H simply yields a difference between the stored energy and that returned to the magnetizing circuit, which is seen to be given by the area under the other half of the hysteresis loop. Thus the net amount of energy loss per unit volume due to cyclic hysteresis may be found by finding the area of the hysteresis loop. The hysteresis loop for a ferromagnetic material may of course be found experimentally, the results plotted and the area of the loop computed. The area will then be a measure of the hysteresis loss.

Hysteresis loss represents an important design factor in electrical machinery. Thus, although hysteresis may be considered an undesirable characteristic of magnetic materials and certain substances such as silicon steel have been developed to minimize this effect, nevertheless, permanent magnets which are of importance in some magnetic circuits is composed of materials with very marked hysteresis effects. Also, most of the self-excited direct current shunt machines, motors and generators, would not be able to build up their voltages if some hysteresis did not occur, with a resulting residual flux in the air-gap to start the build-up process.

When a ferromagnetic material is subjected to an alternating

magnetizing force, the material will be carried repeatedly
through the cycles of magnetization and energy will be dissipated
due to hysteresis loss. At any particular maximum flux density
the loss per cycle will be found from the area of the hysteresis
loop. The power loss, that is, the energy loss per unit time, will
be a function of the number of cycles through which the magnetization is carried. Thus, the power loss or the rate of energy
loss, is equal to the hysteresis loss per cycle times the number
of cycles per second. In transformers the flux is made to vary
between certain definite limits, the iron of the transformer core
is subjected to core losses due to hysteresis. The power loss
will also be a function of the frequency. Here the core loss due
to hysteresis in the iron of the transformer will double, for the
same maximum flux density, if the frequency of flux variation is
doubled. Hence in transformers hysteresis loss becomes very important at high frequencies. Similarly, high speed electrical machinery is more concerned with hysteresis loss than low speed
machinery.

An empiracle equation was developed by Steinmetz, based
upon extensive experimentation, which indicates how hysteresis
loss varies with the maximum flux density which is used.
For alternating magnetizing forces, hysteresis loss depends on
the volume of iron used, the frequency of alternation and the
maximum flux density. Thus the loss for a particular sample of
iron per cubic meter per cycle is constant for a given flux density. Steinmetz therefore sought a practical answer to an important engineering question, namely the dependence of the
hysteresis loss on the maximum flux density. He formulated
his results in a relationship which is quite good for engineering
purposes, if used over a moderate range of maximum flux
densities normally found in electrical power apparatus and for
the types of steels he used. His findings may be summarized by
stating that the hysteresis loss is proportional to the 1.6 power
of the maximum flux density. Steinmetz's equation is

$$P_h = \eta f\, B_{max}^{1.6} \quad \text{joules per cubic meter per second}$$

where P_h is the loss per cubic meter of material per second,
B_{max} is the maximum flux density in webers per square meter,
f is the number of cycles per second of the magnetizing force and
η (eta) is a coefficient whose value depends on the type of material used. Eta (η), known as the Steinmetz coefficient of hysteresis
loss, has values ranging from about 1200 for soft steel to 18,000
for hard tool steel in the MKS system of units. The exponent,
1.6, is quite accurate for maximum flux densities about one

weber per square meter for most materials. For higher values of flux densities, the exponent is not constant and may become as large as 2.5. An example of the application of Steinmetz's equation will prove informative. It is desired to operate a particular transformer at double the normally applied voltage. Experimentally it has been determined that at the normally applied voltage the hysteresis loss is 600 watts. It is also known that if the voltage is doubled the maximum flux density in the transformer core is also practically doubled. If the frequency remains unchanged it is desired to know what the new hysteresis loss will be. Steinmetz's equation predicts that the loss will vary with the 1.6 power of the flux density and since the flux density is doubled then the new power loss will be

$$P_h = 600 \times \frac{(2 \, B_{max})^{1.6}}{(B_{max})^{1.6}}$$

$$= 1815 \quad \text{watts.}$$

Thus, doubling the flux density will result in tripling the loss due to hysteresis.

7.5e Eddy Currents

The Faraday electromagnetic equation states that whenever magnetic flux changes while linking with a coil an emf is induced in the coil; that is, the requirement for electromagnetic induction is that the flux linkages with the closed electric circuit change. The generated emf will produce a current in the closed circuit whose value will be determined by the magnitude of the emf and the circuit resistance. We have already seen that current flowing in a circuit is accompanied by heating losses in the conducting material. Any changes in flux is therefore accompanied by losses in the conducting circuits with which the flux links. The circuit which links with the flux is not necessarily limited to one made of wire nor is it necessary that all the flux penetrate it. When a changing flux passes through a solid mass of metal, such as iron, metallic circuits in the block itself, if they are linked by the flux will have currents flowing in them. Thus it is possible to induce an emf over incremental lengths of the metal which produces currents in the conducting material. The configuration of the conduction paths which are provided determines the manner in which these currents are set up. In large cross-sections of metals the current paths are circular or swirling and they are known as eddy currents. The currents are also signified as stray currents and will, of course, produce losses in resistive materials, such as iron. Such losses are termed eddy current losses. When the changing magnetic flux densities are high, as those normally associated with ferrous materials, eddy currents will produce considerable power losses in the iron unless they are restricted. Taking into account the relation of the direction of the magnetic flux density and the current, eddy currents are always set up in planes perpendicular to the flux. Therefore, in order to minimize the eddy current losses in the iron, instead of using solid material for the magnetic circuit, the iron is subdivided into thin sheets which are stacked parallel to the flux and insulated from one another. When an iron core is built up from such laminations, the circulating paths in which the eddy currents flow are broken up and the eddy current losses are materially reduced. The armatures of direct current generators are built of sheet steel laminations, usually between 0.014 to 0.030 inches thick. These laminations are stacked on a shaft and placed parallel to the working flux of the machine. The high resistance between the laminations as well as the shorter effective lengths of the conducting material at right angles to the flux greatly reduce the magnitude of the eddy currents. For ordinary direct current machines, there is no need to insulate the laminations since the natural oxide on the surface of the sheet steel, after it is rolled and from which the laminations are punched, effectively insulates them from each other, thereby limiting each eddy current path to a single lamination. The laminations of transformers are generally given a coat of shellac or varnish to insulate them. Transformers and machines at high frequencies require extra special care to reduce the eddy current losses for the practical operation of these electrical devices. Thus in generators operating at frequencies of 50,000 cycles per second the thickness of the laminations is kept to about 0.0015 inches and very special precautions are taken to assure adequate insulation between the sheets. Transformers used in communication circuits are sometimes "potted" by using a material of powdered iron and resin, pressed into shape under high pressures, this material having very low eddy current loss.

An approximate analysis, assuming perfect insulation between the laminations, yields an equation from which the magnitude of the eddy current loss may be predicted. Such an analysis gives the following equation for the eddy current loss in iron laminations

$$P_e = \frac{\pi}{6\rho}(h f B_{max})^2$$

where

P_e = eddy current loss in watts per cubic meter

ρ = resistivity of the iron in ohm-meters

h = the thickness of the lamination in meters

f = the frequency of the alternating flux, cycles/second

B_{max} = the maximum flux density in webers/square meter.

From this relation we can see that the eddy current loss is proportional to the square of the frequency and also as the square of the maximum flux density. Thus if either the frequency or the maximum flux density is doubled the loss will be four times as great. The loss also varies inversely with resistivity of the material; the lower the resistivity the higher the loss. The loss is also seen to vary as the square of the thickness of the laminations. A transformer with 0.028 inches laminations of sheet steel will have four times the loss for one of the same volume of iron but with laminations of 0.014 inches of the same grade of steel.

The equation assumes perfect insulation between the laminations and also that the flux density is uniform through out each cross-section of the laminations. It is therefore not surprising that the losses in an actual core may turn out to be twice the value computed from the equation.

7.5f Total Core Losses

We have seen that a changing magnetic flux will produce losses due to cylic hysteresis as well as resistive loss in conducting material due to induced eddy currents. The sum of the two losses, eddy current and hysteresis are called the core loss of the material. We found that hysteresis loss varies directly with frequency while eddy current loss varies with the square of the frequency. Hysteresis loss is proportional to the 1.6 power of the maximum flux density while eddy current loss is proportional to the square of the flux density. These facts are generally stated in terms of the total core loss as

$$P_{e+h} = K_1 f B_{max}^{1.6} + K_2 f^2 B_{max}^2$$

where P_{e+h} is the total core loss and K_1 and K_2 are coefficients, determined experimentally and dependent upon the material used.

The equation enables us to predict the variation of the core loss of machinery with either frequency or flux density if the changes in frequency or flux density are not too large. The total core loss in a machine may be found by measuring the power loss when the machine is operating at no load and only the core losses are present. From the total core loss measurement, it is possible to separate the two losses and determine how much is due to hysteresis and how much to eddy currents. This is usually done by measuring the total core loss at two different frequencies or two different flux densities. This results in two equations which may solved simultaneously for the two core losses. An example will be used to demonstrate this process.

The total core loss of a transformer is measured at 30 cycles and at 60 cycles per second. The flux density at 60 cycles is maintained at the same value as at 30 cycles by proper adjustment of the transformer voltage. It is desired to know the separate losses at 60 cycles. The losses at 30 cycles are 1500 watts, at 60 cycles 3500 watts. By maintaining the flux density constant in making the measurements at both frequencies we may include it with the constant and use the total core loss equation in the following form

$$P_{e+h} = K_3 f + K_4 f^2$$

Therefore, from the measurements we have:

$$1500 = 30 K_3 + 900 K_4$$

$$3500 = 60 K_3 + 3600 K_4$$

The simultaneous solution of the two equations yields:

$$K_3 = 41.6$$

$$K_4 = 0.277$$

Therefore at 60 cycles the hysteresis loss is

$$P_h = K_3 f = (41.6)(60) = 2496 \quad \text{watts}$$

The eddy current loss at 60 cycles is

$$P_e = K_4 f^2 = 0.277 \ (60)^2 = 1000 \quad \text{watts.}$$

Engineers who design electrical machinery usually determine core loss from the core loss curves for the material to be used, which is customarily provided by the supplier. These curves are used not only for convenience but also because the exponent for B_{max} is not always 1.6 in determining the hysteresis loss and because the insulation between the laminations is not ideal. Thus at very low and very high flux densities experimental data, in the form of curves, are more reliable then formulas based on ideal conditions.

CHAPTER 8—MAGNETIC CIRCUIT COMPUTATION

A magnetic circuit usually consists of a high permeability material throughout most of its length, having substantially a uniform cross-section and most of the flux confined to the material. We have seen in Chapter 7 that the reluctance of ferrous materials is usually so much lower than that of the air that surrounds them, that the flux at any cross-section of the ferrous core is very nearly the same. A magnetic circuit is generally designed in order to establish a predetermined amount of flux in a particular space, such as in the air gap between the armature and poles of a generator or motor, using as few ampere-turns as possible. Therefore, the magnetic portions of many types of electrical equipment, such as motors, generators and transformers consist essentially of an iron core which may have narrow air gaps through its cross-section. The engineering problem therefore, becomes one of determining the magnetic flux density or the ampere turns of the magnetomotive force in any particular magnetic circuit. The magnetic circuit calculation is complicated by two significant factors. First, since there is no magnetic insulation comparable to the electrical insulation, leakage flux will always be present. This factor makes it difficult to compute accurately the flux path dimensions. In many applications we may neglect the leakage flux completely and in others simplifying approximations based on experience will adequately account for magnetic leakage. The second difficult arises from the non linearity of the ferromagnetic portions of the magnetic circuit requiring graphical means for the solution of magnetic circuit problems or the so called "cut-and-try" methods.

The most common type of magnetic circuit, found in practice, consists of an iron core, and possibly an air gap, with the magnetizing winding concentrated over a small portion of the circuit. The computation of magnetic circuits usually involve:
(a) The determination of the number of ampere-turns of magnetomotive force that are necessary to set up a desired flux in a magnetic circuit of known geometric form.
(b) The determination of the flux which is produced in a particular circuit by a specified number of ampere-turns of magnetomotive force.

The solution of the first type is usually straight forward and without any difficulty. The second type requires graphical analy-sis or "cut-and-try" methods. It is best to illustrate these com-putations by simple examples. Leakage fluxes are neglected in both cases.

8.1 Ampere-Turns to Produce a Given Flux

The simplest type of magnetic circuit computation is to find the ampere-turns that are required to produce a given flux in a magnetic circuit of uniform cross-section and the same mater-ial throughout. Consider the magnetic circuit of a ring of iron of uniform cross-section whose outside diameter is 7 inches and inside diameter is 5 inches. The ring is made of soft cast steel and circular cross-section. It is desired to find the mmf to set up a flux of 0.0006 weber. Since the flux is establish-ed in and confined to the iron ring whose cross-section is uni-form, then the flux density is readily found and is uniform throughout the entire length of the ring. With the flux density determined we may now find the magnetizing force, ampere-turns per unit length, from the appropriate magnetization curve (B-H curve). We then proceed to find the mean length of the flux path and multiplying this length by the value of the magnetizing force per unit length found from the B-H curve we obtain the necessary mmf.

Since the magnetization curve is given in English units we first convert the flux into kilolines. Thus

$$\phi = 0.0006 \times 10^5 = 60 \quad \text{kilolines}$$

The cross-sectional area, from Figure 44, is

$$A = \pi r^2 = \pi \left[\frac{1}{2} \left(\frac{7-5}{2} \right) \right]^2 = 0.785 \quad \text{square inches}$$

The flux density $\quad B = \dfrac{\phi}{A} = \dfrac{60}{0.785} = 76.4 \dfrac{\text{kilolines}}{\text{square inch}}$

76.4 kilolines per square inch corresponds to a magnetizing force of 31.5 ampere-turn per inch on the B-H curve for soft steel casting in Figure 41.

The mean length of the flux path, from Figure 44 is

$$\ell = \pi \times 6 = 18.8 \quad \text{inches}$$

Since the mmf must be balanced by the total magnetic potential drop in the magnetic circuit then

$$F = H\ell \quad \text{ampere-turns}$$

$$= 31.5 \times 18.8 = 592 \quad \text{ampere-turns}$$

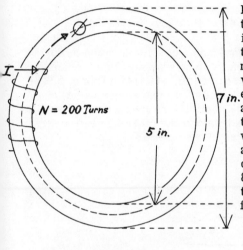

If the coil had 200 turns, the magnetizing current is 2.96 amperes. For this simple type of magnetic circuit the inverse problem is just as easily solved. For example if it is desired to find the flux that would be set up by 800 ampere-turns we can obtain H by dividing 800 ampere-turns by the mean length of the flux path, and obtain

$$H = \frac{800}{18.8} = 42.5 \text{ ampere turns/inch}$$

Figure 44

From the B-H curve for cast soft steel the flux density is

B = 85.5 kilolines per square inch for 42.5 ampere turns per inch.

The total flux set up by the 800 ampere-turns in this simple magnetic circuit will be

$$\phi = BA$$

$$= 85.5 \times 0.785$$

$$= 67 \text{ kilolines or } 0.00067 \text{ webers}$$

8.2 Ampere-Turns to Produce a Given Flux in Magnetic Circuits in Series

Magnetic circuits may comprise two or more distinct parts providing a series path for the flux. The cross-sectional areas may be different for the different parts resulting in different

flux densities in each part. We therefore consider the flux
paths individually. For a given flux, the required mmf may be
found by computing the flux density, the corresponding magnetizing force, from the appropriate B-H curve, the mean length
of the flux path for each part of the magnetic circuit, the $H\ell$
product for that part. The resultant mmf will be the sum of
these products:

$$F = NI = H_1 \ell_1 + H_2 \ell_2 + \ldots$$

Example: Consider the transformer core, with a non uniform
cross-section shown in Figure 45. We wish to find the number
of ampere turns that are required to establish a flux of 0.0006
weber. We may divide this circuit into two series parts; one
consisting of very nearly 42 centimeters long with a cross-
section of 4 square centimeters and the other 20 centimeters
long and 6 square centimeter cross-section.

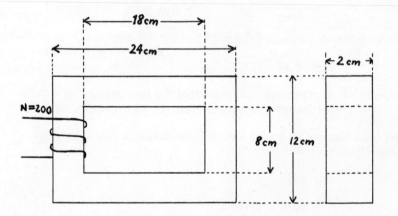

Figure 45

Changing the dimensions to inches, since the B-H curves shown
in Figure 41 are in English units, we have one magnetic circuit with a mean length of flux path of 16.5 inches and a cross-
sectional area of 0.62 square inches and a series circuit with a

mean length of 7.9 inches and 0.93 square inches cross-section. The flux of 0.0006 webers is

$$\phi = 0.0006 \times 10^5 \quad \text{kilolines} = 60 \quad \text{kilolines}$$

We will assume that the core is annealed sheet steel laminations. For a flux of 60 kilolines in the magnetic circuit the flux density in the first part is $\dfrac{60}{0.62}$ or 96.8 kilolines per square inch and in the second part $\dfrac{60}{0.93}$ or 64.8 kilolines per square inch. From the B-H curve for annealed sheet steel the corresponding values of H are 28 ampere-turns per inch and 4 ampere-turns per inch respectively. The magnetic potential drop for the first part of the magnetic circuit is the product of the magnetizing force by the mean length of the flux path of that circuit, i.e.,

$$H_1 \, \ell_1 \; = 28 \times 16.5$$

$$= 463 \quad \text{ampere-turns.}$$

The magnetic potential drop for the second part of the circuit is similarly,

$$H_2 \, \ell_2 \; = 7.9 \times 4$$

$$= 32 \quad \text{ampere-turns}$$

The mmf must be balanced by the total magnetic potential drop or

$$F = \; H_1 \, \ell_1 + H_2 \, \ell_2$$

$$= \; 495 \quad \text{ampere-turns}$$

Therefore to establish a flux of 0.0006 webers in the transformer core requires an mmf of 495 ampere-turns or 2.48 amperes in the 200 turns.

8.3 The Flux Produced by a Given Number of Ampere-Turns in a Series Magnetic Circuit

This type of problem, which involves the computation of the flux produced by a given mmf is simple for a magnetic circuit of uniform cross-section, as was shown before. However, when the magnetic circuit is not uniform, it becomes necessary to use trial and error or the so called "cut-and-try" method. We shall illustrate the method with an example.

We once more consider the transformer core of annealed sheet steel laminations shown in Figure 45. This time we wish to determine the flux which will be set up in this magnetic circuit for a magnetomotive force of 1500 ampere-turns in the concentrated coil. We procede to subdivide the magnetic circuit as we did before into two parts, the first approximately 42 centimeters long with a 4 square centimeters cross-section, the other 20 centimeter long and 6 square centimeters in cross-section. Converting to English units we have the first part of the series magnetic circuit with a mean length of the flux path 16.5 inches and a cross-section of 0.62 square inches and the second part with a mean length of the flux path 7.9 inches and 0.93 square inch cross-section.

We now make a guess at the total flux that will be produced, say 70 kilolines. This guess will enable us to compute the flux densities for each part of the magnetic circuit, and by means of the appropriate B-H curves, the corresponding magnetizing forces. These magnetizing forces are then multiplied by the respective mean lengths of flux path along which we have assumed that they are constant, yielding the magnetic potential drop for each part of the magnetic circuit. The total magnetic potential drop should be equal to the mmf. Usually our first guess will not check but it will give us an indication of a more appropriate guess for our second assumption.

We will now apply this method to illustrate the process. With our first assumption of 70 kilolines for the flux, the flux density in the first part of our magnetic circuit becomes $\dfrac{70}{0.62}$ or 113 kilolines per square inch and from the B-H curve for annealed sheet steel the corresponding value of the magnetizing force, H, is 175 ampere turns per square inch. The flux density in the second part of our circuit becomes $\dfrac{70}{0.93}$ or 75.5 kilolines per square inch for which the corresponding value of H is, from the B-H curve, 6.3 ampere turns per inch. Therefore, the total magnetic potential drop in the magnetic circuit is

$$H_1 \ \ell_1 + H_2 \ \ell_2 = (175) \ (16.5) + (6.3) \ (7.9)$$

$$= 2880 + 49.5$$

$$= 2929.5 \ \ \text{ampere-turns}$$

Our estimate of 70 kilolines is much too high since such a flux requires nearly 3000 ampere-turns and we have only 1500 available. We therefore revise our first estimate. We have learned however that the greatest magnetic potential drop is along the 16.5 inch mean length and that moreover the B-H curve in the region near 113 kilolines per square inch is extremely flat so that a small reduction in flux density means a very large reduction in magnetizing force. We therefore make a more appropriate guess of 66 kilolines for the flux for our second estimate. Then

$$B_1 = \frac{66}{0.62} = 106 \ \text{kilolines per square inch and the}$$

corresponding H_1 from the B-H curve is $H_1 = 80$ ampere-turns.

$$B_2 = \frac{66}{0.93} = 71 \ \text{kilolines per square inch and the correspond-}$$

ing H_2 from the B-H curve is $H_2 = 5.5$ ampere turns.

$$F = NI = 80 \ \text{x} \ 16.5 + 7.9 \ \text{x} \ 5$$

$$= 1320 + 39.5$$

$$= 1359.5 \ \ \text{ampere-turns}$$

Our available ampere turns is 1500 and a flux of 66 kilolines requres 1360 ampere-turns. We now try 66.5 kilolines for the flux with a new value for $B_1 = 107$ kilolines per square inch and a corresponding magnetizing force of $H_1 = 90$ ampere-turns per inch resulting in a magnetic potential drop of $90 \ \text{x} \ 16.5 = 1480$ ampere-turns. The value B_2 is now 71.5 kilolines per square inch and a corresponding magnetizing force of $H_2 = 5.1$ ampere-turns per inch. Thus the total magnetic potential drop is

$$N I = H_1\, \ell_1 + H_2\, \ell_2$$

$$= 1480 + 40 = 1520 \text{ ampere-turns}$$

Thus the required flux is greater than 66 kilolines but less than 66.5 kilolines and very nearly equal to 66.5 kilolines. For engineering purposes the answer of 66.5 kilolines is quite good especially since the hysteresis in the iron will cause an uncertainty greater than the difference between 66 and 66.5 kilolines.

In any such trial and error method it is advisable to keep an orderly solution in order to avoid retracing any steps that may have gone before. A tabular form for the results of the calculation by the "cut-and-try" method will greatly facilitate the computation by having an orderly outline of the various steps in the solution.

Try No.	ϕ	B_1	B_2	H_1	H_2	$(NI)_1$	$(NI)_2$	Total (NI)
		$\dfrac{\phi}{0.62}$	$\dfrac{\phi}{0.93}$	B-H Curve		$16.5\,H_1$	$7.9\,H_2$	
1	70	113	75.5	175	6.3	2880	49.5	2929.5
2	66	106	71	80	5.0	1320	39.5	1360
3	66.5	107	71.5	90	5.5	1480	40.0	1520

We may summarize the "cut-and-try" method for the series type magnetic circuit as follows:

1. Assume a total flux. Although the value of flux selected at the first trial depends a great deal on experience it is advisable not to assume a flux so high that the flux density in the part of the circuit with the smallest cross-section is impossible. Also, if one part of the circuit, such as an air gap, will require more ampere-turns than all the other parts of the circuit combined, we may consider a solution as if this part were the only circuit in the problem.

2. Compute the flux densities for each part of the circuit based on the assumed value of flux.

3. From the B-H curves for the materials used find the corresponding magnetizing force.

4. Find the magnetic potential drop for each part of the circuit by multiplying the magnetizing force for each part by its cor-

responding mean length of path. Add these potential drops to
find the total magnetic potential drop for the entire series circuit.
5. The total magnetic potential drop is then equated to the mmf
required to set up the assumed flux. This mmf is now compared
with the available mmf and a new estimate of the flux is made
based upon the difference between the computed and available
ampere-turns.
6. The sequence is repeated with the revised estimate of the
flux.

8.4 Computations Involving Air Gaps

Electrical machinery, such as motors and generators, have
rotating structures and in order to allow these parts to move
the magnetic circuits include air gaps. In other magnetic circuits air gaps are included to obtain special characteristics.
The calculation of the mmf to produce a given flux in the air
gap or the flux produced by a given mmf is somewhat simplified
by the fact that the permeability is contant for the air gap.
Hence, it is not necessary to use the B-H curve, since the calculation may be made directly from the relation

$$B_{air} = \mu_0 H_{air}$$

To calculate the number of ampere-turns required to produce a given flux across an air gap, we may first compute the
reluctance of the air gap and then use the relation that

$$\phi = \frac{F}{R}$$

from which mmf, F, can be readily found from the product
of the reluctance and the flux, i.e.

$$F = \phi R \quad \text{ampere-turns}$$

Or, we may calculate the flux density, B, in the air gap, divide
B by μ_0 to obtain the magnetizing force in the gap and then find
required mmf by multiplying H by the mean length of the flux
path in the air. Thus, if a magnetic circuit contains an air gap
of 0.08 inches long and an effective cross-sectional area of 3
square inches, we can readily find the necessary ampere turns
to produce 300 kilolines across the gap. All our computations
will be made in the MKS system of units. Thus, we find the re-

luctance from the defining equation

$$R = \frac{\ell}{\mu_0 A}$$

$$R = \frac{\left(\frac{0.08}{39.37}\right)}{(4\pi \times 10^{-7}) \times \dfrac{3}{(39.37)^2}} \quad \text{since 1 meter = 39.37 inches and } \mu_0 = 4\pi \times 10^{-7}$$

$$= 8.45 \times 10^5 \quad \text{ampere-turns per weber}$$

$$\phi = 300,000 \times 10^8$$

$$= 300 \times 10^{-5} \quad \text{webers}$$

Therefore, the mmf is

$$F = \phi R = 8.45 \times 10^5 \times 300 \times 10^{-5}$$

$$= 2535 \quad \text{ampere-turns}$$

We might also have used the second method, namely:

$$B = \frac{\phi}{A} = \frac{300 \times 10^{-5}}{\dfrac{3}{(39.37)^2}}$$

$$= 1.56 \text{ webers per square meter}$$

whence

$$H = \frac{B}{\mu_0} = \frac{1.56}{4 \times 10^{-7}}$$

$$= 1.24 \times 10^6 \quad \text{amperes per meter}$$

Hence

$$F = N I = H\ell$$

$$= 1.24 \times 10^6 \times \frac{0.08}{39.37}$$

$$= 2535 \quad \text{ampere turns}$$

Since the air gap forms a part of the magnetic circuit, the ampere-turns for the air gap must be added to the ampere-turns
for the other parts of the magnetic circuit to find the total mmf
that is required.

It will be noted that we used an effective value for the cross-
sectional area of the gap. The correct cross-section to use for
the computation involving an air gap is generally a matter of
doubt because the flux in the air gap spreads out, and it becomes
necessary to account for this "fringing" effect. The flux lines
upon leaving the iron face of the gap tend to bulge outward and
increase the effective cross-sectional area of the gap, as is
shown in Figure 46. Where the air gap is short, we find that it is
sufficiently accurate to account for the "fringing" by increasing
each dimension used to compute the cross-sectional area by the
length of the gap. Thus in Figure 46, if the length of the air gap
is ℓ_g, the effective cross-sectional area would be considered

$$A = (a + \ell_g)(b + \ell_g)$$

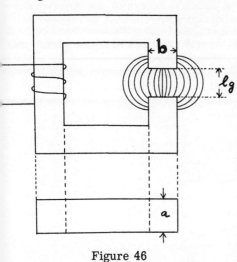

where a and b are the
cross-sectional dimen-
sions of the actual core
faces. If one of the faces
of the gap has a cross-
sectional dimension much
larger than the corres-
ponding dimensions of the
other face, a correction
of $2\,\ell_g$ should be used to
find the effective area of
the gap. Experience shows
that satisfactory results
are obtained by means of
these corrections pro-
vided the correction ap-
plied is less than one-

Figure 46

fifth of the cross-sectional dimension to which it is applied,
i.e., $\ell_g < 0.2a$ or $0.2b$.

For short air gaps with non parallel faces, in chamferred
edges, teeth or other complicating shapes generally found in
electrical machinery, we may resort to semi-empirical rules
for finding equivalent lengths and areas. The method of computing the reluctance of many different sort of air gaps is given by
Herbert C. Roters in his book, entitled, Electromagnetic
Devices.

Leakage flux may be found by dividing the mmf across the leakage path by the reluctance of the path. This reluctance is usually difficult to determine and is generally estimated and checked experimentally.

8.5 Magnetic Circuits with Series-Parallel Branches

We have seen that a single source of emf may produce current flow in any number of parallel branches to which it may be connected. Similarly, a mmf, produced by a single coil carrying magnetizing current, may establish flux in several parallel branches of a magnetic circuit. The core of a three phase shell type transformer is a typical magnetic circuit which is frequently employed. When the winding is placed on the center leg, as is shown in Figure 47, the other two legs serve as a return path for the flux. Magnetic flux lines are always continuous, having no beginning and no end. Hence in Figure 46, the flux, ϕ_b, set up in the center leg must subdivide itself into the two fluxes ϕ_a and ϕ_c set up in the outside legs.

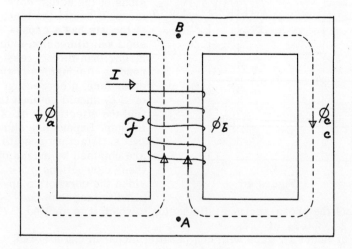

Figure 47

Thus $$\phi_b = \phi_a + \phi_c$$ Eq. 8.51

and in general the flux entering any space in a magnetic field must equal the flux leaving that space in order to satisfy the continuity of the magnetic flux. This relation is very similar to the Kirchhoff current law which we used at current junction points in the electrical circuit calculations.

We see that in order to establish a given amount of flux in the center leg enough ampere-turns must be available to yield the required magnetizing force in the center leg and in either outside leg to satisfy the following relations:

$$F = H_a \ell_a + H_b \ell_b \qquad \text{Eq. 8.52}$$

$$F = H_c \ell_c + H_b \ell_b \qquad \text{Eq. 8.53}$$

That is, the magnetic potential difference between any two points on the magnetic circuit will be the same, no matter what path we may choose from one point to the other provided we include the mmf of the coil as a magnetic potential difference. Thus the magnetic potential drop from B to A in the magnetic circuit of Figure 47 is

$$F - H_b \ell_b \text{ or } H_a \ell_a \text{ or } H_c \ell_c$$

Sometimes the symbol U, is used for the magnetic potential difference. Thus the magnetic potential drop from B to A may be written as

$$U_{BA} = F - H_b \ell_b$$

$$= H_a \ell_a$$

$$= H_c \ell_c$$

where ℓ_a represents the distance from B to A measured along the mean flux path of ϕ_a and ℓ_c is the distance from B to A along the mean flux path of ϕ_c. Thus, equations 8.52 and 8.53 follow from taking this magnetic potential drop along the two paths from B to A

$$F - H_b \ell_b = H_a \ell_a$$

$$F - H_b \ell_b = H_c \ell_c$$

We can see the similarity to the electric circuit shown in Figure 48

Here we see that $I_b = I_a + I_c$. The potential drop from B to A may be written as

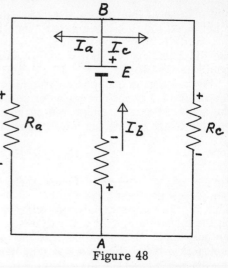

$$V_{BA} = I_a R_a$$

$$= I_c R_c$$

$$= E - I_b R_b$$

Hence

$$E = I_b R_b + I_a R_a$$

and

$$E = I_b R_b + I_c R_c$$

Figure 48

The simplest type of magnetic circuit with parallel branches is one in which the outside legs are exactly alike so that we can assume that the flux ϕ_b divides equally between the legs, so that

$$\phi_a = \phi_c = \frac{\phi_b}{2}$$

We shall illustrate the computation by an example. The dimensions of the magnetic core for a certain transformer are indicated in Figure 47:

Over-all: 20 inches by 12 inches

Each window: 4 inches square

Core thickness: 2 inches

Core material: Annealed sheet steel laminations

We desire to find the mmf supplied to the center leg to establish a flux of 0.00856 weber. In considering the mean lengths of the flux paths we take these lengths as extending from B to A, midway of the core sections.

$$\ell_a = \ell_c = 24 \text{ inches}$$

$$\ell_b = 8 \text{ inches}$$

$$A_a = A_b = A_c = 8 \text{ square inches}$$

Hence $B_b = \dfrac{856}{8} = 107$ kilolines per square inch and

H_b = 90 ampere-turns per inch from the B-H curve for annealed sheet steel.

From the geometry of the core

$$\phi_a = \phi_c = \phi_{b/2} = 414 \text{ kilolines}$$

$$B_a = B_c = \frac{414}{8} = 51.8 \text{ kilolines per square inch.}$$

Hence $H_a = H_c = 2.5$ ampere-turns per inch
Therefore the total mmf will be

$$F = H_a\, \ell_a + H_b\, \ell_b$$

$$= (2.5 \times 24) + (90 \times 8)$$

$$= 770 \quad \text{ampere-turns}$$

For our second example we shall assume that the magnetic circuit of Figure 47 has an air gap of 0.1 inch long in leg c. We now wish to know how many ampere-turns are necessary to establish the same flux 0.00856 webers in leg b.

Because the air gap requires an mmf of $H_{air}\,\ell$ air, the magnetic potential drop from B to A is now modified so that

$$U_{BA} = F - H_b\, \ell_b$$

$$= H_a\, \ell_a$$

$$= H_c\, \ell_c + H_{air}\, \ell_{air}$$

or $$F = H_b\, \ell_b + H_c\, \ell_c + H_{air}\, \ell_{air}$$

The flux density in leg b is still 107 kilolines per square inch, as before, and the corresponding magnetizing force is 90 ampere turns per inch. Since the reluctance of the air gap is large compared to the reluctance of the iron, the flux in leg c will decrease. Since we now do not know how the flux, ϕ_b, will divide we must assume a value and use a trial and error method. We therefore assume $\phi_c = 214$ kilolines. Hence

$$B_c = \frac{214}{8} = 26.6 \quad \text{kilolines per square inch}$$

The corresponding value of the magnetizing force, from the B-H curve is, $H_c = 1.3$ ampere-turns per inch. The magnetic potential drop in the iron for leg c is

$$N I = H_c \, \ell_c = (1.3) \, (24)$$

$$= 31.2 \quad \text{ampere-turns}$$

The reluctance for the air gap is now found

$$R = \frac{\ell}{\mu_o A} = \frac{\dfrac{0.1}{39.37}}{(4 \quad x \quad 10^{-7}) \, \dfrac{(2.1) \, (4.1)}{(39.37)^2}}$$

$$= 3.62 \text{ x } 10^5 \quad \text{ampere-turns per weber.}$$

"Fringing" has been corrected for in the effective area used for the air gap cross-section. The magnetic potential drop across the gap is therefore

$$F = \phi_{air} \, R = 214 \text{ x } 10^{-5} \text{ x } 3.62 \text{ x } 10^{-5}$$

$$= 775 \quad \text{ampere-turns}$$

But the magnetic potential drop from B to A is also

$$H_a \, \ell_a = H_c \, \ell_c + H_{air} \, \ell_{air}$$

$$= 31.2 + 775$$

$$= 806.2 \quad \text{ampere-turns}$$

Therefore $\quad H_a = \dfrac{806.2}{24} = 33.6 \quad \text{ampere-turns per inch}$

The corresponding value of flux density from the B-H curve is

$$B_a = 79 \quad \text{kilolines per square inch}$$

and $\qquad \phi_a = B_a\ A_a = 79 \times 8$

$$= 632 \quad \text{kilolines}$$

We now check our assumption

$$\phi_b = \phi_a + \phi_c = 632 + 214$$

$$= 846 \quad \text{kilolines}$$

And the required flux is 856 kilolines
Our first assumption was a judicious guess, namely, that only one fourth of the flux would be set up in leg c with the air gap as compared to the flux without the air gap. If our estimate had been too high or too low we would now revise it and proceed as before. Our answer is satisfactory from an engineering point of view.

The ampere-turns required are therefore

$$F = H_a\ \ell_a + H_b\ \ell_b$$

$$= 806.2 + (90 \times 8)$$

$$= 1526.2 \quad \text{ampere-turns}$$

Comparing our new mmf with a $\frac{1}{10}$ inch air gap, we can see

that the total length of the magnetic circuit in leg c is 24 inches and required only 31 ampere-turns to maintain the desired flux. Yet, the air gap with a total length of 1/10 inch, as compared with 24 inches, required 775 ampere turns or 25 times as many ampere-turns to maintain the same flux, and justifies the assumption that in the "cut-and-try" method solution for a magnetic circuit with an air gap, a first estimate of the flux density may be found by considering the magnetomotive force to be applied to the air gap alone.

PROBLEMS

Chapter 1

1.1 The free electrons in a metal are in random and chaotic motion, with extremely high individual velocities. Electronic current flow, however, represents the number of coulombs of electricity crossing a particular surface each second. Thus current flow is a transport of charge. A copper wire, 0.162 inches in diameter, is carrying a direct current of 50 amperes. It has been estimated that there are very nearly 8.5×10^{28} free electrons in each cubic meter of the metal. Find the velocity of the drift of the electrons along the wire when the current is flowing.

1.2 The electron-volt is a very convenient unit of energy and is generally in use in electron ballistics and optics, as for example, in the design of the electron microscope. The electron-volt is equal to the amount of energy transferred when an electron is moved or falls through a potential difference of one volt. Determine the number of Joules of energy there are in one electron-volt.

1.3 An electric heater takes 500 watts from 120 volt source. What power would it absorb from a 230 volt source?

1.4 An electric toaster absorbs 1.440 kilowatts when it is connected to a particular line voltage. When this voltage is increased by twenty per cent a current of 10 amperes is found to be flowing through the toaster. What was the original voltage on the line?

1.5 A certain combination of reducing gears and an electric motor constitute a gearmotor. Find the output torque of a gearmotor which has an output of 50 horsepower at a speed of 33 rpm.

1.6 The speed of the electrical motor of the gearmotor, of problem 1-5, is 1800 rpm. If the efficiency of the gearing is 85%, find the torque at the motor shaft.

1.7 For the gear motor of problem 1.5, find the output in watts for the gearmotor and the motor. Find out how much power is lost and what happens to the power which is lost.

1.8 A thermo-couple is a device which converts heat energy into electrical energy. A certain thermo-couple has an emf of 0.0075 volts. At what rate is heat energy being changed to the electrical form when a current of 0.0005 amperes is flowing in the circuit. Compare the temperature of the thermo-couple with the temperature of its surroundings.

1.9 In Figure 1.9, 1.1 kilowatt is converted into electrical energy at terminals a-b of the circuit. Determine the voltages at the other three terminal pairs. Place the correct polarity marks at each terminal pair.

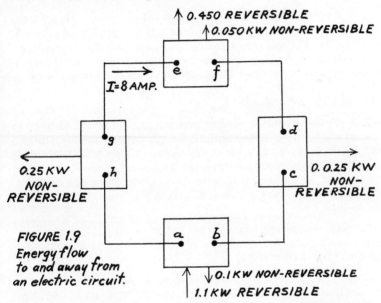

Figure 1.9 Energy flow to and away from an electrical circuit.

1.10 In Fig 1.9, the reversibility of energy flow is shown. Find the following emf's; E_{ab}, E_{cd}, E_{ef}, E_{gh}, E_{gb} and E_{hb}.

Chapter 2

2.1 An electrical heater requires a current of 5 amperes in order to operate at its rated power. If the heater is rated at 1000 watts, find the resistance of the heater. What voltage is required to operate the heater at its rated power?

2.2 An incandescent lamp is rated at 75 watts and 115 volts. Its rated life is 500 hours. Compute the number of Joules of energy that are necessary to operate the lamp for its rated life. What will be the resistance of the lamp when it is being operated at its rated voltage?

2.3 An emf of 208 volts is connected in series with a total resistance 218 ohms. Part of the 218 ohms is a series street lamp, which has a resistance of 20.8 ohms when the current is flowing through the total resistance. Find the voltage drop across the lamp.

2.4 A battery having an emf of 6.3 volts and an internal resistance of 0.01 ohm is being charged by a generator whose emf is 7.5 volts. The resistance of the armature circuit, including the brushes, is 0.05 ohms. The resistance of the interconnecting leads from the generator to the battery is 0.003 ohms. Draw the circuit and determine the potential difference across the terminals of the battery.

2.5 A potentiometer circuit that draws no current is used to measure the emf of a dry cell. The emf, determined by this method, is 1.56 volts. When a voltmeter, whose total resistance is 1000 ohms, is placed across the terminals of the dry cell, the reading on the meter is 1.50 volts. Does this mean that the voltmeter is inaccurate? How do you explain the reading of the voltmeter?

2.6 If a voltmeter, whose total resistance is 100 ohms, is connected across the terminals of the dry cell of problem 2.5 what will the meter read?

2.7 Three fixed resistors, 10 ohms, 20 ohms and 50 ohms are available in the laboratory. By connecting them in various ways, it is not necessary of course to use all three at once, a minimum and maximum value of resistance may be obtained. What are the largest and smallest resistances possible? Tabulate all the resistance values that are possible between these limits. Draw the diagrams for the various connections.

2.8 A slide-wire rheostat, connected as in Figure 2.8, is used as a three-point rheostat to allow the voltage V_3 to be varied from zero to a maximum. If the generator emf is 120 volts and R_3 is 640 ohms find the voltage V_3. What is the power dissipated in each part of the circuit? What is the voltage V_3 when R_3 is changed to 64 ohms?

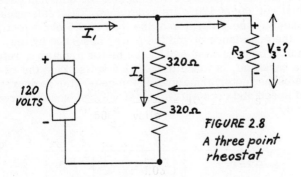

Figure 2.8 A three-point rheostat

2.9 Data for a volt-ampere characteristic was obtained in the laboratory for a 40 watt tungsten filament incandescent lamp.

Voltage across the lamp	Current through the lamp
Volts	Milliamperes
10	120
25	163
40	198
55	230
60	240
75	265
90	290
115	325

Make a graph of the volt-ampere characteristic. Plot voltage (along the ordinate) as a function of the current, (along the abscissa). On the same graph also plot the resistance as a function of the current. From the curves determine the voltage across the lamp when the current is 220 milliamperes. If it is desired to increase this current by 25 per cent by what percentage must the voltage be increased?

2.10 Figure 2.10 represents a d.c. generator which is supplying a load comprising a group of lamps which draw 12 amperes and a motor in parallel with the lamp bank, which draws 12 amperes armature current and one ampere shunt field current. The generator has an armature resistance of 0.1 ohm and develops a terminal voltage of 120 volts under load. The resistance of the line connecting the generator to the load is 0.5 ohms total. Determine the resistance of and the power absorbed by the lamp bank. The generated emf of the motor armature, referred to as the back emf, is found to be 101.5 volts. Find the resistance of the motor armature. If the generator is operated at constant speed and constant flux what voltage would it produce at no load?

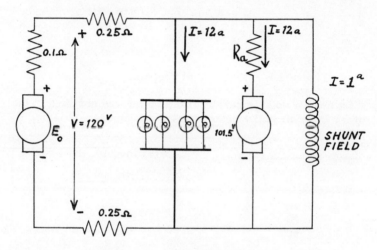

Figure 2.10
D.C. Generator supplying motor and lamp bank load

Chapter 3

3.1 The starter motor of an automobile is operated from a storage battery whose emf is 6.2 volts and whose internal resistance is 0.02 ohms. The armature of the motor has a resistance of 0.008 ohms and develops no "back emf" at standstill. The battery is connected to the starter motor by a footswitch and interconnecting leads whose combined resistance is 0.003 ohms. Neglecting any inductive effects, find the instantaneous current which is supplied by the battery of the starting motor after the foot switch is pressed down.

3.2 Two storage batteries are being charged in parallel from generator which is separately excited. The armature of the generator has an emf of 120 volts and an internal resistance of

0.08 ohms, including the brushes. The emf's of the batteries at their present state of charge are 110 and 112 volts and their corresponding internal resistances are 0.30 and 0.35 ohms respectively. Find the initial terminal voltage of the generator. Compute the initial charging current for each battery.

3.3 A battery which generates 72 volts and whose internal resistance is 0.28 ohms is connected to a pair of terminals x and y by two wires whose resistance is 0.01 ohms for each wire. A load consisting of three resistors in parallel is connected across the terminals x and y. If the load resistors are 10, 18 and 14 ohms respectively find (a) the power dissipated in the 14 ohm resistor; (b) the current in each resistor; (c) the total current taken by the load; (d) the potential across the battery terminals.

3.4 A drop wire is very often used in communication circuits to provide several different voltages and currents as is shown in Figure 3.4.

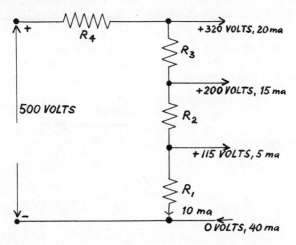

Figure 3.4 A drop-wire for a communication circuit

(a) Find the values of the resistors R_1, R_2, R_3 and R_4
(b) Assuming that the 115 volt tap becomes disconnected but the currents at the other taps do not change find the current flowing through R_4 and the changes that take place in the voltages at the 4 taps.

3.5 An elementary telegraph circuit consists of a transmitter which develops an emf of 100 volts and an equivalent series
resistance equal to 50 ohms. The transmitter in connected to
the receiver by means of a transmission line. The receiver has
a resistance of 50 ohms and requires a minimum current of
35 milliamperes for proper operation. If the transmission line
is 10,000 miles long (forward and return) and has a resistance
of 0.04 ohms per 1000 feet, will the receiver operate satisfactorily. Explain your answer in terms of the maximum allowable
line resistance.

3.6 Three conductances of 0.005, 0.010 and 0.008 mho respectively are connected together in parallel across a potential
difference of 600 volts. Determine the current in the line,
the current in each conductance, the power supplied by the emf
and the power consumed by the 0.010 mho conductance branch.

3.7 An emf of 100 volts has an internal resistance of 0.03
ohms. It is connected to a load consisting of two parallel conductances 0.005 and 0.003 mhos respectively. Find the potential difference across the load and the power consumed by the
0.003 mho conductance.

3.8 For the circuit of Figure 3.8 determine the voltage drop
across the 10 ohm resistor, each branch current, and the total
power supplied by the battery.

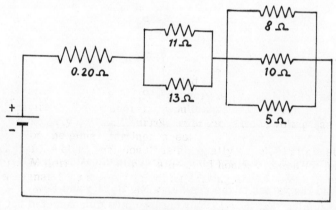

FIGURE 3.8

3.9 In Figure 3.9 load X requires 10 amperes and load Y 8
amperes. Find the voltage drop across each load; the power
consumed by each load, the current in the middle wire.

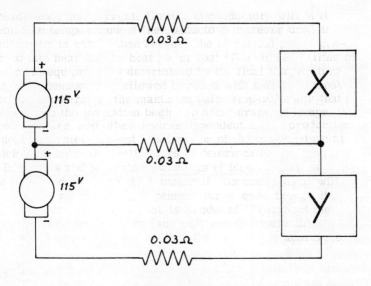

Figure 3.9

3.10 In Figure 3.10 find the currents in each branch.

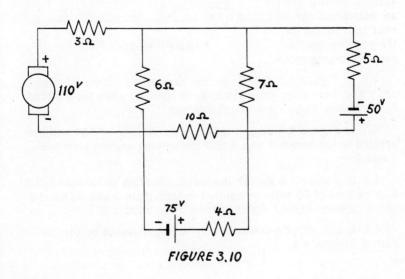

FIGURE 3.10

Chapter 4

4.1 Repeat problem 3.10 using the loop-current method.

4.2 In problem 3.10 convert all the voltage sources to their equivalent current sources and find the power dissipated in the 10 ohm resistor by the nodal-voltage method.

4.3 In the circuit of Figure 4.3 find the voltage across the 25 ohm resistor. In this circuit reduce the delta to an equivalent wye before solving for the voltage.

4.4 If in Fig. 3.9 branch **X** is 0.5 ohms and branch **Y** is 0.8 ohms find the currents flowing in these branches by means of the loop-currents method. What is the current in the middle wire?

4.5 In problem 3.7 replace the voltage source by an equivalent current source and do the problem by the node-voltage method.

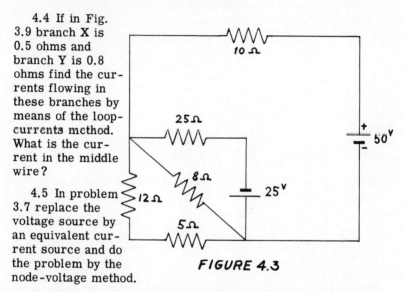

FIGURE 4.3

4.6 In Figure 4.6 determine the resistance between terminals 1 and 2 (a) when the switch, S, is open (b) when the switch, S, is closed. Figure 4.6 is on page 201.

4.7 In Figure 4.6 switch is left open. An emf of 100 volts is applied to terminals 1 and 2 find the voltage across terminals 3 and 4.

4.8 In Figure 4.6 switch is closed, shorting terminals 3 and 4. If an emf of 60 volts is applied to terminals 1 and 2 find the short circuit current flowing through the switch, S.

4.9 A d.c. shunt generator may be represented by the circuit of Figure 4.9

The armature may be looked upon as an emf in series with a resistance, the shunt field may be considered as a resistor and the load is a resistance load. If the armature resistance is 0.28 ohms, the field resistance 123 ohms and the load requires 5 kilowatts at 115 volts determine:

(a) the emf of the armature

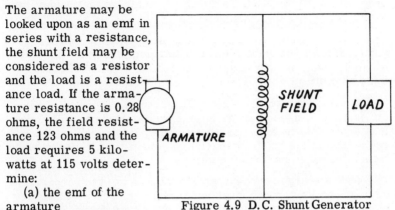

Figure 4.9 D.C. Shunt Generator

(b) the current flowing in the armature

(c) the power which is converted to electrical form in the armature.

4.10 Do problem 3.2 by the nodal-voltage method.

Chapter 5

5.1 Find the branch currents in problem 3.10 by means of the Superposition Theorem.

5.2 Do problem 4.4 by the Superposition Theorem.

5.3 Find the current in the 10 ohm resistor of problem 4.3 by means of Thevenin's Theorem.

5.4 In the circuit of Figure 5.4 determine the value of R_L for the maximum power transfer. Using the value of R_L find the maximum power which is transferred to the load. Also find the ratio of the power output to the power input.

5.5 If R_L is made equal to 50 ohms in Figure 5.4 find the voltage

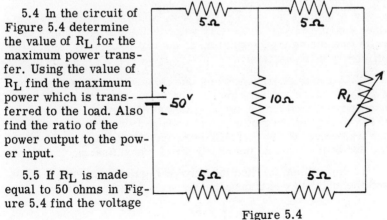

Figure 5.4

drop across R_L by Norton's Theorem; also find the ratio of the output current to the input current.

5.6 A d.c. generator delivers 8.5 kw to a load which consumes 8.3 kw. The total resistance of the distribution line which connects the generator to the load is 0.158 ohms. Determine the voltage at the generator and also at the load.

5.7 An energy source consists of a constant current of 150 milliamperes, shunted by an internal conductance of 0.015 mho. It is desired to transfer maximum power to a load. Find the resistance of the load; the maximum power; the efficiency of the transmission at maximum power.

5.8 The energy source of problem 5.7 is used to supply power to a load R_L through a two wire transmission line whose resistance is 72 ohms. Find the value of the load resistance for maximum power transfer, find the maximum power and the efficiency of the system.

5.8 In problem 3.3 convert the voltage source to an equivalent current source and find the potential across the battery terminals and the current through the 10 ohm resistance by Norton's Theorem.

5.9 Repeat problem 3.3 by Thevenins' Theorem.

5.10 Do problem 3.2 by the Norton Theorem. Would there be any simplification if the voltage sources are first replaced by their equivalent current sources before the Norton Theorem is applied? Explain.

Chapter 6

6.1 A 50 kw load is energized from a 750 feet tramsmission line; the line voltage drop is 17 volts. If the load voltage is 240 volts find the voltage regulation of the transmission line and the efficiency of transmission.

6.2 A two wire transmission line is used to supply power to a 440 volt direct current motor. The motor is rated at 30 horsepower shaft output and operates at 85 per cent efficiency. If the voltage at the motor is 440 volts and each conductor of the transmission line is 0.013 ohms find the voltage at the sending end of the line and the efficiency of transmission.

6.3 In problem 3.9 find the voltage regulation at each load and the efficiency of transmission.

6.4 An electric toaster consumes 1.5 kw when it is directly connected to a 120 volt d.c. supply. We will assume that the resistance of the toaster does not change. What power will a 120 volt d.c. source supply to the toaster through 150 feet of No. 10 AWG wire. Find the efficiency of transmission and the voltage regulation at the toaster.

6.5 Determine the probable wire size for a two wire copper distribution line which is used to connect a d.c. generator, whose terminal voltage is 230 volts, to a d.c. motor which delivers 15 horsepower at an efficiency of 84 per cent. The voltage at the motor is 220 volts and the motor is 500 feet from the generator.

6.6 A particular type of standard annealed copper wire has a diameter of 257.6 mils. Find the resistance of 1000 feet of the wire at 40°C.

6.7 A steel street car rail has a resistivity of 115 ohm-circular mils per foot at its normal operating temperature. The cross-sectional area of a single rail is 2.8 square inches. Compute the resistance of 3 miles of track consisting of two rails in parallel.

6.8 The field coil of a generator is wound from commercial annealed copper wire. Its resistance at 17.5°C is 25 ohms. After the generator has been operating for a long period of time the resistance of the field coil is found to be 34.5 ohms. Compute the average temperature rise of the field coil.

6.9 The resistance of the field coil of a particular loud speaker is 8 ohms. The field coil is rated at 500 milliamperes but requires a minimum current of 385 milliampers for satisfactory operation. The speaker, placed at the front of an auditorium, is 200 feet away from the 6.3 volt power supply. Will a pair of AWG No. 14 copper conductors provide satisfactory operation. What is the maximum distance from the voltage supply to the speaker that will produce satisfactory operation if AWG No. 12 copper wires are used.

6.10 A voltage supply is located one half mile from a 125 kw load. The voltage at the load is 750 volts. If the transmission efficiency is 93 per cent what is the smallest commercial size of annealed copper wire that can be used, the operating temperature is 20°C.

Chapter 7

7.1 A cast-iron ring, of square cross-section, has an inside diameter of 5 inches and is 0.6 inches thick. This core is wound with a coil of 500 turns. A current of 1.5 amperes flowing in this coil sets up a flux of 0.95×10^{-4} webers. Determine the average magnetizing force and the permeability of the cast iron.

7.2 A soft-steel core has a relative permeability of 2500. The core has a square cross-section of 7 square inches and is 30 inches long. When this magnetic circuit is energized by 72 ampere-turns find:
(a) The reluctance of the magnetic circuit
(b) The flux set up in the core
(c) The magnetizing force
(d) The magnetic flux density.

7.3 A cost-iron toroidal core has a cross-sectional area of 22×10^{-4} meters and a mean length of 1.2 meters. A coil of 300 turns carrying 5 amperes is wound uniformly on this core. Find (a) the mmf; (b) the flux set up in the core, (c) the magnetic flux density; (d) the magnetizing force and (e) the reluctance of the magnetic circuit.

7.4 It is desired to produce a flux of 0.02 webers in a soft-steel ring whose cross-sectional area is 5 square inches and mean circumference is 20 inches. A coil of 100 turns is uniformly wound on this ring. Find (a) the current required to produce the desired flux and the resultant mmf.

7.5 In problem 7.4 find also the magnetizing force in ampere-turns per inch and in ampere turns per meter; the magnetic flux density in webers per square meters and also in kilolines per square inch. What is the relative permeability of the steel ring at the conditions under which it is operating?

7.6 A magnetic circuit which is constructed from sheet steel laminations is found to have a loss of 50 watts at a frequency of 60 cycles and a flux density of 0.75 webers per square meter. At a flux density of 0.38 webers per square meter at the same frequency the loss is measured at 32 watts. Find the eddy-current and hysteresis losses at each flux density.

7.7 A magnetic circuit is found to have a loss of 47 watts at 400 cycles and a flux density of 0.2 webers per square meters. If the flux density is maintained at the same value and the loss at 60 cycles is found to be 38 watts determine the eddy-current and hysteresis losses at this flux density.

7.8 A particular sample of iron, a rod of 0.25 inches in dia-meter and 6 inches in length, is subjected to a cyclic magnetiza-tion and a hysteresis loop is obtained and plotted on cross-sec-tion paper with scales of one inch equivalent to 0.5 webers per square meter and 800 ampere turns per meter. The area is measured, when so plotted, and found to be 9.5 square inches. Find the energy loss represented by the area of the hysteresis loop in Joules per cubic inch per cycle.

7.9 A transformer core has a volume of 14 cubic inches. The hysteresis loss at 30 cycles is measured as 3.5 watts for a maxi-mum flux density of 50 kilolines per square inch. Find the coefficient of hysteresis loss. What will be the hysteresis loss at 25 kilolines per square inch and a frequency of 60 cycles per second?

7.10 Determine the range in value of the coefficient of hystere-sis loss in Steinmetz's equation for ordinary annealed sheet steels that are found to have hysteresis losses ranging from 1.0 to 2.0 watts per pound at a flux density of 64.5 kilolines per square inch and at a frequency of 60 cycles. The specific gravity of the steel is 7.5.

Chapter 8

8.1 In the magnetic circuit of Figure 8.1 what current is re-quired to set up a flux of 0.0009 webers? Assume that the core is made of annealed sheet-steel laminations.

8.2 A magnetic circuit is made of an-nealed sheet steel. The circuit has a uniform cross sec-tion of 4.2 square in-ches and a mean length of 2.5 feet. Determine the num-ber of ampere turns to set up a flux of 0.006 webers in this circuit? It is desired to decrease the flux in the core by 25 per cent, by how much should the current be increased or decreased?

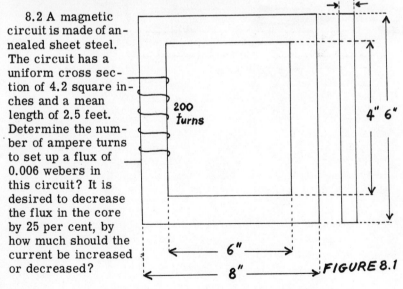

200 turns

1.125"

4" 6"

6"

8"

FIGURE 8.1

8.3 In the magnetic circuit of Figure 8.1 an air gap is introduced in the middle of one of the shorter legs by cutting a space of 2 millimeters wide. Find the current that will now be required to produce a flux of 0.0009 webers through the magnetic circuit.

8.4 In Figure 8.4 the magnetizing coil carries 5 amperes and has 200 turns. The core consists of annealed sheet steel. The paths A and C are each 35 centimeters long and leg B is 15 centimeters long. The cross-section is uniform throughout the entire magnetic circuit and is 12 square centimeters. Find the flux density in each part of the circuit.

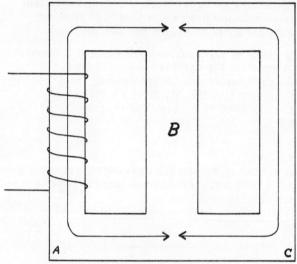

FIGURE 8.4

8.5 In the magnetic circuit of Figure 8.1 an air gap of 0.05 millimeter is made in one of the longer legs and a current 0.40 amperes flow in the coil. Find the total flux produced in the magnetic circuit.

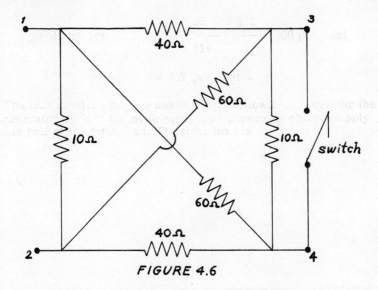

FIGURE 4.6